Lessons amid the Rubble

JOHNS HOPKINS INTRODUCTORY STUDIES
IN THE HISTORY OF TECHNOLOGY

Lessons amid the Rubble

An Introduction to Post-Disaster Engineering and Ethics

SARAH K. A. PFATTEICHER

The Johns Hopkins University Press

Baltimore

Printed in the United States of America on acid-free paper
9 8 7 6 5 4 3 2 1

The Johns Hopkins University Press
2715 North Charles Street
Baltimore, Maryland 21218-4363
www.press.jhu.edu

Library of Congress Cataloging-in-Publication Data

Pfatteicher, Sarah K. A., 1968–
Lessons amid the rubble : an introduction to post-disaster engineering and ethics / Sarah K.A. Pfatteicher.
p. cm. — (Johns Hopkins introductory studies in the history of technology)
Includes bibliographical references and index.
ISBN-13: 978-0-8018-9719-1 (hardcover : alk. paper)
ISBN-10: 0-8018-9719-X (hardcover : alk. paper)
ISBN-13: 978-0-8018-9720-7 (pbk. : alk. paper)
ISBN-10: 0-8018-9720-3 (pbk. : alk. paper)
1. Structural failures. 2. Disasters—Research—Ethical aspects. 3. Structural engineering. 4. Disaster relief. I. Title.
TA656.P488 2010
624.1'71—dc22 2010001220

A catalog record for this book is available from the British Library.

Special discounts are available for bulk purchases of this book. For more information, please contact Special Sales at 410-516-6936 or specialsales@press.jhu.edu.

The Johns Hopkins University Press uses environmentally friendly book materials, including recycled text paper that is composed of at least 30 percent post-consumer waste, whenever possible. All of our book papers are acid-free, and our jackets and covers are printed on paper with recycled content.

Contents

Introduction

Why?

On the second Tuesday of September 2001, the Twin Towers of the World Trade Center returned to dust, taking with them some 2,700 souls. A disaster of such magnitude asks us to explore our beliefs—about religion, surely, but also about politics and social systems, and even about technology. As we search for answers to the deceptively simple question "Why?" we may find ourselves in difficult and troubling territory. As the *New York Times* noted, "The value of the national commission . . . investigating the terrorism of September 11, 2001, is that it forces the nation to look beyond its feelings of grief and awe, and grimly assess the things that could have been done better."[1] It is painful to question the truths we thought we knew. And yet a questioned faith can become stronger for that questioning.

Since the fall of 2001, hundreds of articles, videos, and lectures have attempted to construct from the wreckage an explanation for the devastation. Some (like *Frontline*'s "Faith and Doubt at Ground Zero") have sought a spiritual explanation for (and resolution of) the disaster. Some (like Yosri Fouda and Nick Fielding's *Masterminds of Terror*) have explored the terrorists' political motivations and the social conditions that helped them to grow. Still others (most notably, perhaps, the FEMA *Building Performance Study*) have taken the eminently practical approach of focusing on physical explanations.[2]

Whatever their approaches, many of these explorations have

been inspired by four common questions: Why did it happen? Did it have to happen? Who was responsible? What can we learn from it? These questions are not unique to engineering, nor to this particular crisis, but the nature of the attacks gave them a particularly acute form. In the face of disaster, it is natural to ask why the unthinkable has happened: it can form an essential part of the process of grieving and of recovery. In the midst of confusion, it is natural, too, to find comfort in objectivity and neutrality of the sort provided by scientific explanations. We may disagree about what sort of god would allow so many to die, and we may debate the sociopolitical systems that nurtured the terrorists. But perhaps there is some common ground and even solace to be found in reconstructing the mechanical sequence of the collapses; perhaps we can, at least, prevent a similar collapse from happening in the future. And so it has been that engineers have taken on the task of explaining the devastation of that fall day.[3]

As forensic engineers dug through the debris of the Twin Towers looking for lessons amid the rubble, many professional engineering societies used the turn of the millennium as an impetus to reexamine their missions, their assumptions, and their futures. Organizations such as the American Society of Civil Engineers (ASCE), the National Academy of Engineering (NAE), the National Council of Examiners for Engineering and Surveying (NCEES), the Accreditation Board for Engineering and Technology (now called simply ABET), and the American Society for Engineering Education (ASEE) have been engaging in a dialogue about the future of the profession. In NAE's case, these discussions have taken the form of exploring "the engineer of 2020." How will the engineering profession change in the coming years? How will funding (for research and design) be altered in the twenty-first century? In what ways will liability be reinterpreted post 9/11? What changes will be necessary in engineering education to respond to this altered landscape?[4]

How might the questions being asked and answered about the collapse of the World Trade Center help us to develop a deeper understanding of engineering more broadly and thus inform discussions about the engineer of 2020? It may seem counter-intuitive to focus on a spectacularly unique terrorist attack in search of fundamental lessons about engineering. I do not mean to imply that engineering is at heart about failure or devastation or death. The truth is that I believe disasters highlight a great deal that is good about the profession. Much the way that death can cause us to stop and consider the value of life, looking carefully at disasters can help us to value the successes that we so often take for granted.

As sociologist Eric Klinenberg found in his "social autopsy" of a deadly heat wave in Chicago, the most influential day-to-day operations of a city, an agency, or a community can be hidden before our eyes until they are brought into bold relief by unexpected and critical events, providing an ideal opportunity for study. As Klinenberg evocatively put it, "Institutions have a tendency to reveal themselves when they are stressed and in crisis." Along similar lines, Steven Biel introduced his volume *American Disasters* by explaining that

> catastrophic disturbances of routine actually tell us a great deal about the "normal" workings of culture, society, and politics. . . . The victims and witnesses of calamitous events can't help but try to make sense of disasters in the terms available and comfortable to them. Established patterns of language, belief, and behavior become a significant kind of reassurance in the face of disaster: assertions of order and continuity against disorder and disruption.

Indeed, if we cast our net broadly enough, catastrophes such as the collapse of the Twin Towers can teach us not only about the construction of buildings, but about the structure of engineering itself.[5]

The audience in my mind as I write this is represented by the diverse group sitting in my Introduction to Engineering Design class. I choose this class because it is here that we as engineering educators set the stage for what's to come—in the curriculum and in the professional practice for which we are preparing our students. The room is filled to overflowing with nearly 300 first-semester engineering students. To them, I want to give a sense of what this field is about and why, in the face of 9/11, it is still worth entering.

Mixed in among these new students are roughly 20 upper-level engineering undergraduates, who are helping to teach the first-years. I want to draw these senior students out of their increasingly narrow coursework to consider the big picture of the profession they are about to enter, and their roles in it. A handful of graduate teaching assistants line the back wall; they come from a variety of fields and have chosen to study the profession of engineering at a deeper, more advanced level. I want to help them reconsider what they think they know about engineering and about design, leaving them, in the end, not more cynical, but more respectful of its power.

Ten faculty and instructors are scattered about the room, keeping watch over their groups of students in this vast lecture hall. Many of them are teaching first-years for the first time, and it is my aim to help them to look at their profession through the eyes of a newcomer and at the same time to explore a more philosophical view of the endeavor we are undertaking. And occasionally, hovering by the doors, there is a visitor to this land of engineering—a prospective student visiting campus, perhaps, or a non-engineer passing time, curious about what engineering is. With them I want to share, if only briefly, a glimpse of the beauty, complexity, import, and appeal of design. These people come to my classroom with various levels of expertise in engineering, but my aim is to inspire and challenge them all with the material I present.

At the start of the fall semester, near the anniversary of that awful day, we study the untimely demise of the World Trade Center. It seems at first a stand-alone topic, inserted as a forced memorial rather than as core material. In short order we move on to cover the steps in the design process, from identifying the client's needs to brainstorming ideas, from researching the state-of-the-art to building models, from testing prototypes to constructing the final product, all the while communicating with peers and clients. What I hope becomes clear to students as the semester progresses is that beyond the hands-on, nitty-gritty feel for design that students and instructors alike gain from Introduction to Design, there is room for broader questions. What ties these topics together—the real world immersion in the projects and the more abstract explorations of the profession—is that they each provide a sense of the nature of engineering for better and for worse.

We begin Introduction to Engineering Design by asking students to define engineering, then spend the semester exploring the meanings and purpose of the field by diving into projects. The essays here encourage readers to step back from day-to-day practice to reflect on the broader meaning of this field. One elegant description of engineering is "design under constraint." This definition reflects just one of the dichotomies that make up this profession: design requires creativity, yet not the unbridled creativity of the artist, for the engineer is confined by physical, legal, financial, and other constraints. Engineering is made up of such contradictions. They make the field at once fascinating and frustrating.

The pages that follow begin with an overview of investigations into the collapse of the Twin Towers, then take half a dozen events connected with the collapse of the World Trade Center and read each for its broader context and significance, ultimately arriving at six deceptively simple lessons about the engineering enterprise:

1. Engineering theory speaks of certitude, but engineering practice is characterized by ambiguity.
2. Engineers strive for perfection, and yet they must expect mistakes.
3. Engineering expertise is often necessary and yet insufficient to solve the problems of the world.
4. Engineers value public safety, but they must balance this ideal with other needs.
5. Engineers succeed through commitment, hard work, and allegiance, but these habits are best when practiced with restraint and moderated by other influences.

6 Finally, engineering curricula tend to reward memorization and conformity, but engineering design requires a balance of creativity and conservatism.

(One might reasonably include a seventh lesson on the tension between engineers' pride in their accomplishments and their disdain at the lack of respect afforded them by the public, but this is a topic for another time and place.)

These characterizations are simply stated, but each embodies an ethical dilemma at the heart of engineering. Together, they demonstrate that ethical and professional responsibility cannot be taught in a stand-alone module any more than one can provide a rich sense of the meaning of "engineering" in a single classroom session. Indeed, issues of professionalism and professional identity already are infused in engineering education, whether we explicitly recognize it or not. The essays here are intended to bring such assumptions to the fore where they can be discussed explicitly rather than hidden in the subtext of a curriculum. In sum, this unpacking of the collapse of the Twin Towers provides a means for exploring the purposes, goals, and responsibilities of the American engineering profession.

Students in introductory physics learn that light is both particle-like and wave-like—and yet is neither. To understand how light behaves, one must hold two seemingly contradictory views, at times emphasizing one or the other, but always remembering that these apparently mutually exclusive characteristics both help define what light is. So too with engineering: it is contradictory, made up of perspectives and practices that seem at first polar opposites and yet operate jointly. In certain circumstances, it is possible to view engineering more simply, but these simplifications cannot capture the full nature of the endeavor. Yes, engineering strives for certainty and perfection and public safety, but it also wrestles with ambiguity and failure and cost constraints.

For physics students, understanding the dual nature of light is difficult but essential. To some extent the duality must be taken on faith and yet continually explored. For engineering students, understanding the dual nature of engineering is similarly challenging and necessary, requiring faith and exploration. The essays here are meant to stimulate that exploration as a means to strengthen the underlying faith.

I find myself endlessly fascinated by the seemingly impossible combination of characteristics engineers exhibit. They are eager to push the boundaries of what we know and can do, while simultaneously cautious and reserved in fostering change. On one level this pairing of disparate characteristics makes sense. How else would engineers have sent men to the moon and brought them back safely, as John F. Kennedy exhorted them to do? How else would the Brooklyn Bridge have spanned the East River with such strength and yet such grace? How else would they have both the boldness and the wariness to develop new biomedical devices, power plants, automobiles, and pesticides?

Of course engineers should possess both daring and caring. What is baffling and intriguing is how the profession manages to find tens of thousands of such individuals each year. In an essay on the collapse of the World Trade Center, Eric Darton remarked that "we humans are born creatures of the earth and air, capable of functioning with our heads in the clouds—so long as our feet remain on the ground." Darton may be right that all humans possess this dual nature to some extent, but psychologists report that the "engineering mindset" is in fact rather rare. According to the Myers-Briggs personality-type schema, the engineering profession is dominated by the rarest of the sixteen personality types, that known as INTJ, for those who possess a preference for Introversion, iNtuition, Thinking, and Judging. INTJ's indeed are the essence of the engineering personality: careful thinkers willing to question assumptions and eager to take action. In other words, if engineers seem unusual, it's because statistically speaking they are.[6]

The approach I use to introduce and explain the profession of engineering is drawn from my interdisciplinary perspective, working in engineering education and teaching engineers and yet not an engineer in any strict sense of the word. I hold a bachelor's degree with a double major in physics and mathematics and earned master's and doctoral degrees in the history of science, with an emphasis on the history of engineering ethics and design. My research and publications have been primarily in two areas: the intersection of engineering disasters and ethics on the one hand, and undergraduate engineering education and assessment on the other. I served, from 2001 through 2007, as assistant dean for academic affairs in the College of Engineering at the University of Wisconsin–Madison and am also a member of the Science and Technology Studies Program at the UW–Madison, a dual position that has afforded me cherished opportunities to participate in and reflect on the nature of this beast we call engineering.

My thoughts in the pages that follow have been stimulated by a variety of provocative readings. I would recommend four of these to any engineering student as essential to developing a sophisticated appreciation for this profession and the obligation its practitioners have to practice it well. Langdon Winner's classic essay "Do Artifacts Have Politics?" challenges us to consider the ways in which technology itself possesses values, apart from those who create and use it, and thereby emphasizes the power inherent in the engineered environment. Michael Martin and Roland Schinzinger's chapter "Engineering as Social Experimentation" pushes the limits of the analogy between engineering and medicine by exploring the obligations engineers have to seek "informed consent" from those who will be testing their work.[7]

Charles Perrow and Eric Darton each make what at first appear to be outlandish claims about the nature of engineering. Perrow's classic work, *Normal Accidents*, argues that failure is an inherent part of modern-day engineering, which is characterized by complex systems; indeed, he argues convincingly that our attempts to reduce risk can in fact increase the likelihood of failure. Darton's more recent essay, "Architectural Terrorism," dares to compare designers' and terrorists' ways of making sense of the world and encourages us to adopt what philosopher Michael Davis has called "thinking like an engineer" with caution.[8]

All of these works challenge the notion that engineering is simply "applied science." Together, they paint a picture of engineering as a complex interaction of people, interests, powers, and objects. If done poorly or with little insight, it is a profession with immense potential to do harm. If done well and with an appreciation for its nuances, it is a profession that truly can do a great deal of good in the world.

I hope that in reading the essays here, readers will know the admiration and affection I have for this profession, even as I challenge assumptions about what engineering means and how engineers should go about their work. As I suggested in the opening paragraph of this introduction, questioning a faith is a difficult undertaking, but one that can strengthen our beliefs and fortify our commitment. We owe as much to the many victims of September 11. May they all, the living and the dead, find peace.

FOR FURTHER EXPLORATION

1. Before reading further, how would you define the following terms?

Engineering	Engineer
Failure	Accident
Design	Problem Solving
Safety	Creativity
Responsible	Professional

2. Do your definitions change if you compare pairs of terms, reading from left to right (e.g., engineering and engineer)?
3. As you continue reading, return to these terms and definitions. Do their meanings change for you or stay consistent?
4. What significance do these terms and their meanings have for engineering education? Why?

CHAPTER ONE

"A Very Imperfect Process"

Engineering Problem-Solving 101

The events of September 11 . . . are not well understood by me . . . and perhaps cannot really be understood by anyone. So I will simply state matters of fact.

LESLIE E. ROBERTSON

As the dust settled and the fires died at Ground Zero, a group of engineers climbed over the piles of debris that had been sent by barge to the Fresh Kills landfill on Staten Island, looking for clues to what had happened on that clear September morning. They had come as part of a hastily convened investigative panel jointly organized by the American Society of Civil Engineers and the Federal Emergency Management Agency. They had little hope of offering solace to the grieving families, and no hope of reversing history, but these engineers intended to put their technical skills to use in helping the country understand a piece of what had happened to one of the great engineering achievements of the century.

Within hours of the collapse, ASCE formed a team to investigate the collapse from an engineering perspective, to explain the sequence of events that resulted in the collapse of both towers. Within a month, FEMA officially became ASCE's partner in the study.[1] The results of that investigation were published in late summer 2002 in a volume blandly titled *World Trade Center Building Performance Study*.[2] Conducted under the direction of Gene Corley (well known and widely respected for his role in investigating the collapse of the Murrah Federal Building in Oklahoma City), the FEMA/ASCE study struggled with problems of minimal funding and limited access to the WTC wreckage.[3] The resulting report provided a factual, technological analysis of what happened during the morning of 9/11, beginning with the moment the planes entered the towers. What proved to be more difficult was to entwine this story of the disaster with other perspectives to achieve a holistic understanding of what had happened on that Tuesday.

From an Engineering Perspective

The study team arrived on site in the midst of a confused and crowded bustle of activity. Recovery efforts were still under way, and authority on the site had not yet been clearly established.[4] Sam Melisi, a firefighter and construction worker who emerged as a leader in the early days after the collapse, described the site as something of a free-for-all: "It wasn't regulated at all. The first couple of days, anything went. It wasn't like somebody was saying, 'You can't go in there, you can't do this, you can't do that.' It was more like 'Hey, if you think you can get in there, go ahead.' "[5] In the midst of grief and terror and the overwhelming task of "unbuilding" the pile the towers had become, few people were thinking of the wreckage as a pile of potential evidence. FBI efforts focused on documenting the actions of the terrorists in the hours, days, and months leading up to the attack. The focus of public and governmental attention was on responding to the attacks, whether through grieving or retribution. As shocking as most of us found the collapse of such huge buildings, the minute mechanical details of how the buildings fell seemed, in those early days, somewhat beside the point. We were, as some said, facing a whole new world, in which life and our security seemed to have changed irrevocably. In the face of such human drama, it would take a particular mindset to view the devastation and say, as officials at ASCE effectively did, "We need to get over there and document what happened to the beams and columns and trusses in that pile."

What drove this engineering team to undertake their analysis? The simple answer is that engineering analysis is what the team members were trained to do. They responded, as many volunteers at Ground Zero had, in the way they knew how, using the tools and skills they possessed. But we might, without engaging in psychoanalysis of the team's motivations, press for additional complexity in our understanding of why the members pursued their investigation as they did. Some engineers clearly believed an engineering analysis would provide the best possible explanation of the collapse. Others clung to engineering as a useful starting point in the midst of so much uncertainty.

In December 2001, two engineers from the Massachusetts Institute of Technology called on nineteenth-century Scottish mathematician and physicist Lord Kelvin to help explain the importance of understanding the mechanics of what happened to the Twin Towers: "I often say . . . that when you can measure what you are speaking about, and express it in numbers, you know something about it; but when you cannot measure it, when you cannot express it in numbers, your knowledge is of a meager and unsatisfactory kind."[6] The question we are most likely to ask in the face of any disaster or major misfortune is "why?" and we tend to respond based on our primary worldview. Samuel Florman has written extensively on "the engineering view" and describes one of its characteristics as "familiarity with science—and with the technological applications of science—[which] helps make us feel at home in the world."[7] Engineers with a deep and abiding faith in the power of numbers would, from this perspective, never be at peace with the collapse of the World Trade Center until it could be analyzed and expressed in quantifiable, mechanical, engineering terms.

If the MIT engineers, like Kelvin, believed true understanding lay in the mathematics and physics of the collapse, there were others who viewed such objectivity as a useful rest stop on the road to understanding, rather than the ultimate destination. For this latter group, there were emotional benefits to be gained from science. A reviewer on Amazon.com praised a documentary on the FEMA/ASCE study for being "emotion-less" and therefore useful in helping her process such an overwhelming event. As Sophie M. "Nessinette" explained,

> After the tragedy, media overflew us with emotional info, and sincerely, I did not turn the TV on for a while. I was pregnant, and just thinking of all the kids who lost their parents made me sick. This documentary helped me to have another view on the events. Very technical, it doesn't forget the value of each life, but focus on what could be (could have been/should have been) done to avoid such tragedy. After viewing it, you can view another doc[umentary], such as the report by the 2 french brothers [the documentary film entitled *9/11*], and then, you can focus on the emotions, because your mind can "understand" the technical.[8]

Leslie Robertson, structural engineer for the World Trade Center, expressed a similar affection for the technical version of events in his "Reflections on the World Trade Center." He had been asked to explain what he could about the collapse. He was willing to try, but cautioned that "the events of September 11 . . . are not well understood by me . . . and perhaps cannot really be understood by anyone. So I will simply state matters of fact."[9] For Nessinette and Robertson, and perhaps for members of the study team as well, focusing on the technical offered a respite from the overwhelming emotion of grief, and a safe harbor protected from the sea of unfathomable questions about how the world had arrived at this point.

The *New York Times* echoed this need to deal with an overpowering event in stages, declaring that "the value of the national commission now investigating the terrorism of Sept. 11, 2001, is that it forces the nation to look beyond its feelings of grief and awe, and grimly assess the things that could have been done better." The *Times* suggested that the engineering response was more productive and more focused on the future than some other responses, noting that "to point out, as former Mayor Rudolph Giuliani did yesterday, that the terrorists were responsible is both accurate and unhelpful." To move the nation from grief to action, from past to future, would require a more complex analysis of the events of September 11, and engineers were well poised to contribute to that analysis.[10]

These three explanations of why the engineers focused on the mechanics of the collapse—that it was where their skills lay, that it was necessary for a complete understanding, and that it offered a manageable response to an unmanageable event—all have some truth to them and can be applied to American engineers' responses to a variety of engineering disasters over the years. But the fact that the collapse of the Twin Towers was precipitated by terrorists added a complication to this investigation. Studying the performance of the towers would open the door to the troubling possibility that terrorists were not solely to blame, suggesting that the engineers and terrorists were in some way complicit. In a nation poised for war and seeking vengeance, the FEMA/ASCE team may well have preferred not to conflate the engineers' role with that of the terrorists. That the attacks came at all was a shock to most Americans. Yet even as this reality sank in on that fateful morning, many of us understood that the ability of the two buildings to absorb the impact of two 767s was remarkable, if temporary. Our amazement turned to horror as the sudden and largely unexpected collapse began. The buildings were, it seemed, both stronger and weaker than we imagined. In light of such a set of surprises, it was logical and appropriate for the study team to begin simply by establishing the facts of the day.[11]

Engineering Problem-Solving

Whatever their individual motivations, the engineering team members gathered in New York to offer their services. They immediately faced a series of practical questions. First, who was in charge? In other words, how would they gain access to the site, and which investigators were primary? Second, where to begin? The pile was vast, and one couldn't save everything. Where was the team going to store the evidence they did want to save? Third, who was going to pay for this investigation? ASCE and FEMA had gathered a modest budget, but most of the investigators' time and efforts would be volunteered to the project. Fourth, what exactly were the goals of their study?

Let's begin, as the team did, with some basics. Engineering problem-solving is an orderly undertaking that can be described as occurring in five phases. (Truth be told, there are as many descriptions of engineering problem-solving as there are engineers, but these descriptions tend to have basic similarities.) Although the phases overlap and fold back on one another, for the purposes of our overview, let's take them in linear order. Once a need has been identified (phase zero, if you will), phase one focuses on establishing factual starting points: what is the scope of the problem and how can this be expressed in specific, quantifiable terms? Phase two moves on to generating and comparing ideas: what possible solutions might solve the problem and which ones are worth pursuing? Phase three involves testing and evaluating the ideas generated in phase two: how well does each potential solution actually meet the specifications laid out in phase one? Phase four requires that decisions be made—about which solution is most appropriate and whether refinements are necessary. Phase five involves communicating the final solution and making recommendations about how the client should move forward.

In solving any engineering problem, each phase relies on and may feed back to those preceding it. Furthermore, though each step is important in its own right, all five must be completed for a problem to be solved, a design to come to fruition, or an investigation to be considered complete. Importantly, this model for problem solving assumes that there is no single right or best answer; rather,

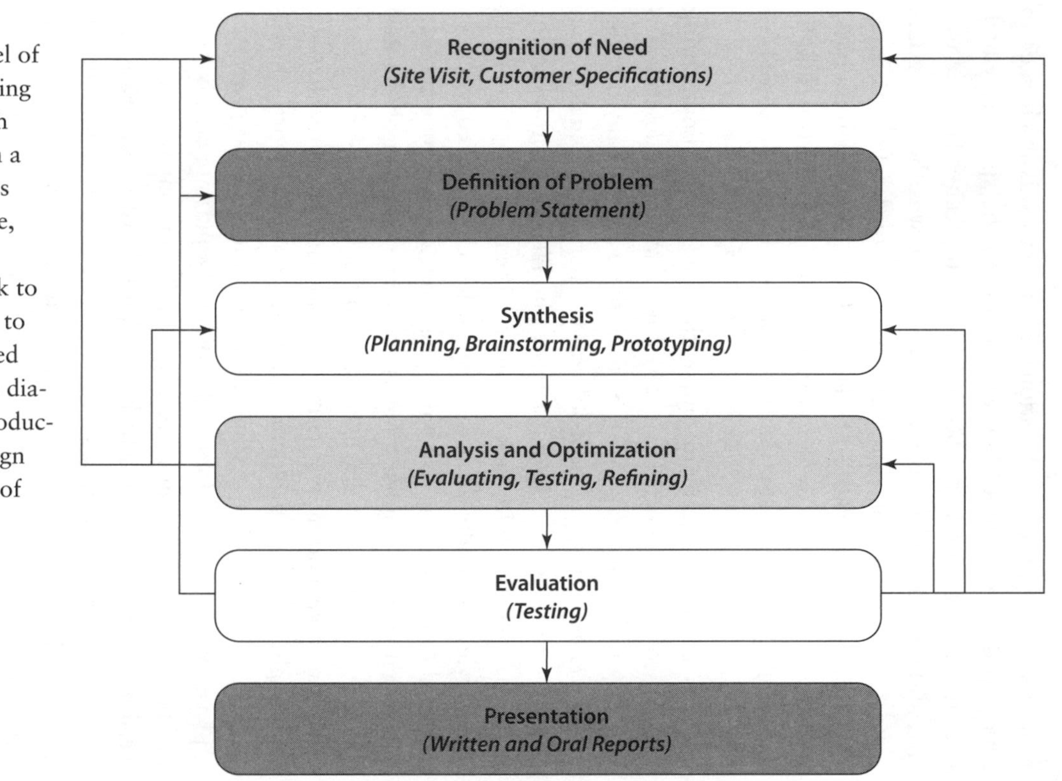

Phases of Design. Flow chart showing one model of the steps of the engineering design process. Although the steps are arranged in a line, the design process is in reality an iterative one, as indicated by the multiple arrows folding back to earlier steps. My thanks to David Hoerr, who created an earlier version of this diagram for use in our Introduction to Engineering Design course at the University of Wisconsin–Madison.

judgment and choice are key elements throughout all five stages. It is worth noting that some engineers suggest that design in fact is far broader than these five stages, beginning well before scoping the problem and proceeding far past stage five to the uses and consequences of the design. We'll return to this point later, but for now it is sufficient to focus on these five core steps.

Skill with this sort of orderly, logical problem-solving is often cited as the most valuable product of an undergraduate engineering education because the steps described here are so broadly applicable. Problems in engineering design, in professional ethics, even in the political arena can be productively addressed using this organized process. Indeed, these problem-solving techniques are useful in helping us to understand the issues involved in the WTC disaster, which in turn will prove useful in elucidating the nature of engineering practice itself.[12]

As powerful as this approach is, it must be applied with caution and an awareness of its limitations. Who decides or defines the problem to be addressed? Who determines which facts are relevant? On what criteria are possible answers evaluated? In short, it is worth asking repeatedly what is missing or obscured by the apparent simplicity and orderliness of such a model.

With these five phases in mind, what might an investigation into the collapse of the Twin Towers look like? We can expect it to begin by establishing the facts of the collapse—what happened and when? Investigators would use those facts to produce theories about why events unfolded as they did, and then test and evaluate those theories until they had compiled the facts into a plausible story arguing whether the buildings performed well or poorly. In stages four and five, the investigators would make an informed, professional judgment about the most likely cause of the collapse, then communicate that judgment and recommendations leading from it to their peers and/or the public.[13]

The FEMA/ASCE study team did indeed progress through these five steps in their investigation. In the dry words of the report itself, "The purpose of this study was to examine the damage caused by these events, collect data, develop an understanding of the response of each affected building, identify the causes of observed behavior, and identify studies that should be performed."[14] Behind this orderly analysis lay two equally important, if unspoken, questions. One, could another building have performed better? And two, what responsibility for the buildings' performance lay with the designers? At each stage of the investigation, these two questions would shape the team's efforts, but neither could be answered quantitatively or simply.

Fact-Finding, or How Bias Enters into a Seemingly Objective Process

Engineering has been elegantly defined as "design under constraint," and indeed, the study team would find their investigative design constrained early and often. When the team arrived at Ground Zero, with just $1 million of the estimated $40 million necessary to do a thorough investigation, they found a great deal of confusion and very little control. For starters, the site was huge—some 17 acres—and the pile was high, rising 150 feet, or five stories tall, with an additional six stories of debris packed underground.[15] Furthermore, the number of people on the scene was mind-boggling, and no one seemed to be in charge of the "unbuilding" process, at least in any official way. As William Langewiesche of the *Atlantic Monthly* described it,

> The mayor's Office of Emergency Management was supposed to be coordinating things, and the main response was overwhelmingly and appropriately the Fire Department's. . . . None of this reflected the normal operation of the U.S. emergency-response system. . . . in the past . . . the efforts on the ground had rapidly been nationalized under the direction of FEMA and its operational allies in the Army Corps of Engineers. . . . The system was intelligently organized and charted.[16]

Although this structure existed in theory, and had worked in the past, in reality neither the OEM nor the FDNY had control of the Trade Center site, and FEMA did not arrive until early October.

Langewiesche continued: "For the first few weeks there must have been two dozen chiefs, each doing good work and believing to various degrees that he or she was the one in charge." Those on the site quickly came to believe that "there was a new social contract here, which everyone seemed to understand. All that counted about anyone was what that person could provide now." As the days progressed, the Department of Design and Construction—"an obscure bureaucracy 1,300 strong whose normal responsibility was to oversee municipal construction contracts"—became the de facto leader on the pile. One DDC official explained: "We just went in and did what we had to do. And no one said no." Asked by Langewiesche if anyone had said yes, the official responded, "No. But then again, we weren't asking."[17]

The FEMA/ASCE team realized as well that their objectives were competing with a variety of other activities at Ground Zero: firefighters searching for survivors, construction crews clearing debris, FBI agents scouring the pile for criminal evidence, the Environmental Protection Agency checking for toxins in the rubble and the air, and other engineers busily "mapping the chaos" to ensure the safety of all these teams on the site.[18] Study team leader Gene Corley understood the need for each of these activities, but argued that the efforts of the forensic engineers needed to be a priority as well.

> Obtaining access to the site of a disaster is always difficult and clearly the search and rescue efforts and any criminal investigation must take first priority. However, in all studies of this nature, gaining access to the site as soon as possible is important in order to observe and document the debris and site conditions. For the future, it may be useful to consider some protocol or process whereby selected individuals from the BPST [Building Performance Study Team] would be allowed on site

in the initial days after a catastrophic event to gather critical data.[19]

In this bustle of activity, amid frantic but fading hope of finding survivors, little thought had been given to salvaging pieces of what one observer described as a gargantuan bundle of steel wool. Despite these obstacles, both physical and organizational, Corley and his team were determined to do what they could to analyze "the response of the buildings, including fire behavior, structural design, fireproofing characteristics, and damage resulting from the aircraft impacts."[20]

The team scattered and began to gather what evidence they could find. They had no clear authorization to do so, but neither did anyone at the site stand much in their way. Some team members remained at the site, watching the unbuilding and marking what pieces of the structure seemed to hold promising clues. Other team members headed to the Staten Island landfill and two recycling yards in New Jersey in hopes of rescuing other components of the building before they were sold for scrap. Still other team members searched out videos and photographs of the towers from the morning of the attack, as well as testimony from eyewitnesses and the original blueprints and building plans.[21]

One of the challenges of the fact-finding stage is knowing how much data is enough. For the study team, this decision was especially vital because most of the physical evidence would be destroyed as it left the site—there would be no going back to retrieve a missed clue. On October 12, the study team wrapped up its initial round of data gathering at the site. It had catalogued 156 pieces of steel, and would view hundreds of hours of videotape and thousands of still photographs, but the evidence remained piecemeal. In part this was due to access and timing, and in part it was due to the physics of what happens to a 110-story building collapsing at a speed approaching 120 miles an hour.[22] As two *New York Times* reporters explained, "The laws of physics hold that energy is not destroyed. All the power used by the construction workers to lift steel, pour concrete, hammer nails had been banked in the buildings as potential energy for three decades. . . . All [278 megawatt hours] of it was released at the moment of the building's demise, floors picking up speed as they slammed downward."[23] That anything at all recognizable remained after such a calamity was due to chance or fortune, depending on your point of view.

From Corley's perspective, one of the key contributions to the investigation would prove to be "the application of methodical science" as much as the evidence to which that science was applied. One journalist reported, "Corley had confidence that the catastrophe could be reasoned through," that this oversized ball of steel wool would yield its secrets to engineering problem solving.[24] And yet, though the team labored long and hard gathering evidence and compiling facts, their investigation had already been shaped by factors out of their control and could not be a purely objective analysis. True, it would be based on facts, but on a particular subset of facts: what photographic and material evidence happened to exist. From the beginning the idealized phases of engineering problem-solving were constrained by reality and the messiness of life—by the chance of what survived, which moments were captured on film, and what survivors could recall. This bias was inevitable rather than intentional, but these constraints nevertheless served to shape the findings even before the investigation had begun. Aware of their handicaps, the study team decided early to offer suggestions for future study rather than put forth concrete answers or proposals, offering their findings as starting rather than finishing points.

It was easy to see how the evidence and happenstance would shape the findings; it was perhaps less clear how the team members' perspective—in particular their amazement that things had not been far worse that morning—would affect their study. There had been some 14,000 individuals in the Twin Towers at 8:45 a.m. on September 11, 2001. That so many of them had time to escape seemed miraculous. The final death toll at the World Trade Center would amount to fewer than 3,000—still a huge loss but so much less than if the buildings had collapsed sooner or toppled instead of pancaking. Gene Corley would admit later that the study team had been fascinated by how long the towers survived. Meanwhile, Trade Center structural engineer Leslie Robertson was privately agonizing over the fact that the towers had not stood "one minute longer." But if Robertson was questioning his role in the collapse, the study team was not: their focus remained firmly on the performance of the building, not that of its designers.[25]

Analysis: When There Is No Precedent, Return to the Basics

Facts in hand, the team turned their efforts to making sense of the data they had acquired. When an engineering team enters the idea-generation phase, a useful place to begin is by asking how others have solved similar problems before. Here, the FEMA/ASCE team faced yet another challenge, for there had been no event quite like this one.

The collapse of the Murrah Federal Building in Oklahoma City in 1995 had also been the result of a terrorist act but had been on a vastly smaller scale, with a building roughly a tenth the size of just one of the Twin Towers, with far fewer deaths and less than total destruction of the building. Gene Corley's experience leading the Oklahoma City investigation would be important, but the parallels were far from perfect. A B-25 airplane had flown into the Empire State Building in 1945, but again, the scope of the damage was nothing compared with what the study team faced in Manhattan in 2001. The largest structures ever to have collapsed were probably those intentionally demolished for development purposes. The Loizeaux family, famed proprietors of Controlled Demolition, Inc., for example, have a renowned understanding of building collapses from their half century of experience in bringing down unwanted bridges, hotels, sports arenas, and more.[26] Each of these cases could provide clues as to where to look for evidence in the Twin Towers, but never had a building of this size been brought to earth in quite this manner.

Thus, there were few useful or perfect comparisons, but engineers drew what they could nonetheless. As it would happen, a more appropriate analogy would be found in the sinking of the *Titanic*, but that realization would take time to emerge. For now, the team would do best by returning to basics.

Why do buildings fall down? Gravity, of course. Which components tend to fail first? Forensic engineers understood that the most vulnerable part of a structure tends to be its connections. In the case of the Twin Towers, had the impact or the fire been primarily responsible for the collapse? Given that the towers did not immediately collapse, attention naturally focused on the effects of the fires. How do materials behave under force or heat? Do they expand, contract, flex, cleave? Such were the questions the team had the ability to answer. The evidence was piecemeal and the lessons of history served as imperfect guideposts, but the team was able to use their knowledge to propose some likely scenarios and suggest avenues for further study.

Through careful analysis of the available wreckage, video footage of the collapses, and blueprints of the buildings' design, the team—described by one observer as "cautious, preliminary, and publicity shy"—pieced together a likely sequence of events for each tower. To understand the team's findings, we must first understand how the towers had been constructed. The WTC designers had opted to forego more traditional framing, in which a tower is essentially a skeleton of stacked layers of cubes, in favor of a strong tube (or "skin") and a compact core joined by a stabilizing set of lightweight floor trusses.[27] The impact of the planes damaged but did not devastate the buildings. Although many of the exterior columns were damaged or destroyed, the loads they had carried were transferred to the remaining columns, in what is known as a Vierendeel truss, an archway of sorts over the devastated sections. The fire, fed initially by copious airplane fuel but within minutes nourished primarily by readily available office paper, weakened the structure and precipitated the collapse.

Connections probably failed first, as the steel trusses tying to-

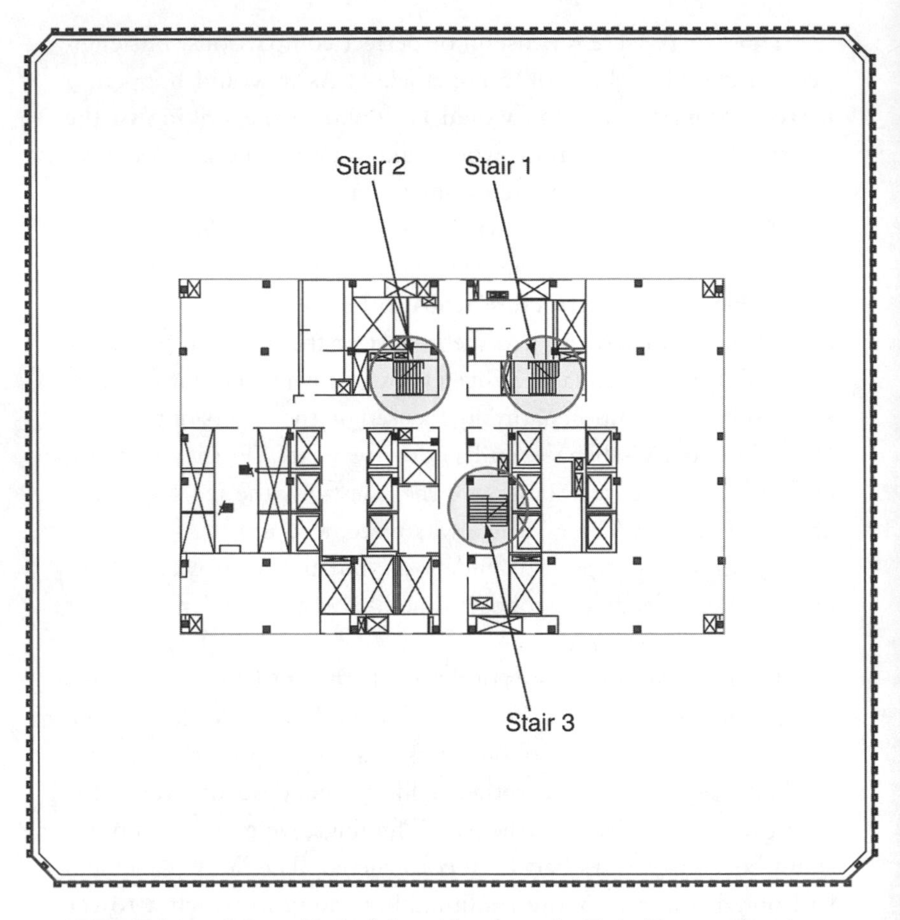

WTC Tube and Core. Typical floor plan from the Twin Towers, showing outer tube and compact core that together provided the key structural stability for each tower. Note the clustered elevator shafts and stairwells within the core and the open space between the core and the outer tube. The location of the stairwells varied from floor to floor rather than running directly up and down each building. At the site of impact, the stairwells in Tower 1 were more closely clustered than those where the plane entered Tower 2. As a result, the plane's entry into Tower 1 severed all evacuation routes for those above the impact, whereas in Tower 2, one stairwell remained intact, allowing some on floors above the impact to escape. ASCE/FEMA, *World Trade Center Building Performance Study*, www.fema.gov/rebuild/mat/wtcstudy.shtm, p. 2-14.

gether the tube and the core of the building sagged and pulled away from the columns to which they were attached. The trusses were particularly vulnerable because fireproofing was either missing (due to remodeling over the years or due to the impact of the planes) or ineffective (given the extreme heat of the fire—it would become clear eventually that the fireproofing specifications had been based largely on guesswork). As the floors gave way and the exterior walls lost their lateral support, the weight of the upper floors took over where the fires left off: gravity pulled them toward earth, in a pancaking progression that gained momentum with each slap of a floor onto those below. The investigators also suggested that the sequence of collapse seemed to differ in the two towers, with the outer tube of the building giving way first in the South Tower, and the core being the first to collapse in the North Tower (the one that sported the massive antenna).[28]

Assessment: Success or Failure Depends on How You Define the Problem

The stages of problem solving are funnel-like. At each stage, the potential endpoints are further constrained or narrowed. The definition of the problem defines the facts to be gathered; the collected facts in turn limit the ideas that can be generated. In gauging the success of their model of the buildings' collapse, the study team based their assessment on the goals they had established from the beginning: they were, after all, conducting a building performance study, not an analysis of the evacuation or emergency management procedures, much less a study of the role of terrorism or international relations in the events of that warm September morning. Dealing with all of these at once would have been a formidable challenge, to be sure. As the team focused on the more manageable question of the building's performance, a troubling question loomed: could another building have stood longer?

The FEMA/ASCE report had stated its purpose as being to

"develop an understanding of the response of each affected building." In an interview, study leader Gene Corley described the investigators' efforts more crisply: "The question that fascinated our team was not, 'Why did the towers fall as quickly as they did?' but, 'Why did they stay standing for so long?' "[29] The team, then, had determined early on that the towers had performed successfully. To what did they attribute this?

The investigators praised the performance of the towers, remarking, "Events of this type, resulting in such substantial damage, are generally not considered in building design, and the ability of these structures to successfully withstand such damage [for as long as they did] is noteworthy."[30] The team went on to explain:

> The ability of the two towers to withstand aircraft impacts without immediate collapse was a direct function of their design and construction characteristics, as was the vulnerability of the two towers to collapse [as] a result of the combined effects of the impacts and ensuing fires. Many buildings with other design and construction characteristics would have been more vulnerable to collapse in these events than the two towers, and few may have been less vulnerable.[31]

In other words, according to the *Building Performance Study*, the towers had performed better than average. William Langewiesche, the one journalist allowed free access to the site during the cleanup stage, detected "an element here of a profession preparing to circle the wagons." He claimed that the study team members "understood the deal" inherent in their task.

> The attack on the World Trade Center was an act of war. Despite the occasional chatter in the press about shoddy steel or substandard fireproofing, the towers were as well designed, built, and maintained as could have reasonably been expected in America in the late twentieth century. . . . The towers fell because they were severely maimed and sprayed with burning jet fuel; they fell as any building will, no matter how resilient, if

it is hit by the next bigger missile in the escalating progression of war.[32]

From Langewiesche's perspective, the team had begun their work with the assumption that the terrorists alone could be blamed for what had happened—a not uncommon viewpoint in those early days and weeks. That the stairwell design had affected the evacuation of the building "was an error that would have to be avoided in future designs. But there was no point in wishful thinking here."[33]

The FEMA/ASCE study team was not alone in defending the WTC design. From the earliest days after the collapse, the towers had been praised for their strength. Just two months after the collapse of the towers, the *New York Times* described the ways in which the towers had played a "lifesaving role," noting, for example, that "the stiffness of the columns kept the buildings from tipping over." Langewiesche, reporting for the *Atlantic Monthly*, marveled "that the building was able to swallow an entire 767, and to slow it from 590 mph to a stop in merely 209 feet." At a gathering of building industry leaders just five weeks after the attacks, one attendee remarked that "the buildings performed heroically, given what was thrown at them." A colleague agreed that "the buildings did perform admirably" and that defending the buildings would be important to public confidence in the months after the attacks. In a discussion of how to respond to the attacks, one engineer suggested that "one of the things that can be done rather quickly is to identify the success stories in these buildings and transmit that information to the general public."[34]

Two *New York Times* reporters explained this emphasis on the towers' success as deriving from the shock of the early days, which instilled in many viewers a tendency to tell and believe stories woven from need as much as from fact. For example, initial reports suggested that firefighters had died valiantly trying to save lives, which seems in retrospect to have been true only in a loose sense, given that the vast majority of WTC occupants had gotten themselves out of the building without assistance and that few firefighters had reached the floors where remaining tenants were trapped. So, too, with early assumptions that the attacks themselves were unanticipated and thus unpreventable, which later gave way to reports that various federal agencies possessed intelligence that could or should have suggested an attack of this sort was imminent.[35]

Dwyer and Flynn from the *New York Times* suggest that a similar set of myths developed around the performance of the towers themselves. They note that the FEMA/ASCE team "concluded that it had been primarily the impact of the planes and, more specifically, the extreme fires that spread in their wake, that had caused the buildings to fall, and nothing that they termed a 'design' flaw." While this version seemed plausible, even comforting, in late fall 2001, the reality was more complex and more troubling: "For decades, the generations that rode higher and higher into the upper floors of skyscrapers had taken it on faith that the evolution of such buildings had been solely a story of progress, of innovations and enhancements that made new buildings safer than the old. The WTC towers were presented as marvels, as buildings so robust they could withstand the impact of an airliner." And yet, although the designers of the towers had considered the potential effects of a plane crash, they had not taken into account what the fuel or fires from such a plane would do to the towers. There had been no analysis of what thickness of fireproofing was required to ensure even two hours of stability for the structural components in a blaze. Although the design team had done groundbreaking studies on tenant reaction to building motion (such as can be expected in a tall tower buffeted by oceanic winds), there had been little apparent concern for how to evacuate tens of thousands of those tenants from 220 acres of space during an emergency.[36]

These facts allowed doubt about the "success" of the buildings to creep in around the edges, and there were individuals who questioned the findings of the *Building Performance Study*. An early *NOVA* documentary on 9/11 asserted: "It is hard to imagine that these are the ruins of buildings so stalwart and strong that they actually saved people's lives. Yet this is the central conclusion of the [FEMA/ASCE] report and its most controversial finding." That same film ends with a comment from structural engineer Charles Thornton, designer of the Petronas Towers in Malaysia: "A lot of people are saying that the structural engineering of the World Trade Center was miraculously wonderful, that the buildings stood up in the case of two 767s flying into it. I would tend to think they were not as successful as they could have been."[37]

But in general, observers carefully avoided any conclusion suggesting that the performance of the towers constituted a "failure" in the usual sense of the word. Perhaps this early faith in the successful performance of the towers, the federal government, and the firefighters is a natural component of looking for good amid the evil that seemed overwhelming in mid-September 2001, but such claims did not arise from an analysis of the facts of the collapse itself. The *New York Times* noted in 2002: "As scientists and engineers have gained . . . hard-won glimpses into the mechanics of the tragedy, there is one other question that almost all of them have carefully avoided asking: could another building, indeed any building, no matter how stoutly or cleverly built, have stood longer than the twin towers did, let more people escape or perhaps never collapsed?"[38] In other words, it is one thing to analyze and explain how the towers collapsed; it is quite a different matter to assess whether they should have been expected to perform differently than they did.

It is true that much of the early coverage of the 11th of September focused on the role of the terrorists, and clearly, looking beyond that requires a certain diplomacy: it is politically difficult and emotionally painful to suggest that anyone but the terrorists might have contributed to the nearly 3,000 deaths that day. As two *New York Times* reporters put it, "Nothing can diminish the culpability of the hijackers and their masters in the murders of September 11, 2001, which stand beyond mitigation as the defining historical truth of the day." Those who dared to suggest that fault lay not just with the terrorists faced swift retribution.[39]

Witness, for example, the reaction to Colorado professor Ward Churchill's description of the WTC victims as "little Eichmanns," effectively ordering their own demise. Churchill lost speaking engagements and was threatened with termination of both his job and his life for his gall. When a New York City task force suggested changes to the city's building code based on what was being learned about how the Twin Towers had performed, one member of the ASCE study team responded angrily that the towers' collapse was not due to their design: "The towers collapsed because they were struck by 767 aircraft used as military-type weapons. *Everything else* was a consequence of that action—*not* a cause. No commercial office building in the world has been designed to survive such an attack."[40]

The notion that there might have been multiple, interwoven reasons for the collapse seemed unacceptable, and yet catastrophic structural failures never come about for simple reasons. In reality, what happened that day is more complex—the towers fell because of the terrorists *in combination with* other factors. So how can the civil engineering profession deal with those other factors? What are they and what roles did they play? What lessons can and will be learned that will help shape the future of design?[41]

Judgment: No Simple or Perfect Solution Exists

If questions about the performance of the towers on 9/11 were difficult to answer, an even more challenging question lurked in the background: What was to be made of the performance of the towers' designers?

The "verdicts" or judgments on the towers' designers have fallen roughly into one of two categories: those praising the WTC engineers for every minute the towers did remain standing, and those who—inadvertently or not—blamed the designers for each minute the towers did not survive. Within the engineering profession, praise for WTC structural engineer Leslie Robertson and his colleagues is widespread, but this praise is not universal. As the study team investigated the performance of the buildings, a documentary film crew set its sights on the performance of the building's designers and drew some disquieting conclusions.

The Learning Channel's *World Trade Center: Anatomy of the Collapse* purports to follow the ongoing investigation, but from the opening music, TLC's video clearly has a dramatic, almost tabloid-like style. The film is framed around two vivid questions: "What was the cause of the devastating, explosive failure that happened to the World Trade Center?" and "How did such a massively strong steel building suddenly give way and come crashing down to earth?"[42]

Anatomy of the Collapse provides stunning computer-generated imagery to help viewers visualize what was happening to the interior structure of the towers while we watched their exteriors on television. But interwoven with these impressive technical displays, *Anatomy* suggests that there was a sinister additional cause for the collapse—one that can be laid at the feet of the towers' designers as well as the terrorists. The voiceover that was originally planned for one shot of the smoldering ruins sums up TLC's analysis of the collapse as follows: "While no one doubts the blame for the tragedy lies squarely with the terrorists who caused it, it now seems that at least some of the loss of life could have been prevented before one of the world's premier buildings was reduced to this." In the final version, the video declares more tactfully that "new evidence suggests that the buildings may have been uniquely vulnerable to the attack and that their design prevented people from escaping before one of the world's most renowned buildings was reduced to this."[43]

Perhaps the most incriminating evidence suggested by TLC comes from Leslie Robertson's own words: "The responsibility for . . . the ultimate strength of the tower was mine, and the fact that they didn't stand longer could be laid at my feet." Asked if he felt remorse about the towers' collapse, a clearly agonized Robertson replied, "Do I feel guilty about it, about the fact that they collapsed? The circumstances on the eleventh of September were outside of that that we considered in the design. Today if I knew then what I know now they would have stood longer of course." This characterization of Robertson as partly to blame is in stark contrast to FEMA's report, which doesn't mention him at all by name, and *NOVA*, which paints a sympathetic picture of a tortured man who did his best and whose crowning achievement was destroyed by terrorists. Compare, for instance, TLC's suggestion that blame for the collapse "could be laid at [Robertson's] feet" with *NOVA*'s words: "For Leslie Robertson . . . [t]here is a special kind of torture: his office overlooks what was once his greatest achievement. 'Ground Zero is a very disturbing place for me. . . . And I cannot escape the people who died there. Even if I'm looking down into a pile of rubble, it's still, to me, somehow up there in the air, burning. And I cannot make that go away.' "[44]

There is no easy answer to the question of whether the towers could have or should have stood one minute longer. One minute bleeds so easily into two minutes, and on into five . . . . How much time would have been necessary to allow all the tenants and firefighters and others to evacuate? It is important to remember that Leslie Robertson and his colleagues were not the root cause of the destruction on the 11th of September. From that perspective, even to question the guilt or innocence of these engineers is hurtful and inappropriate, as they were not in the cockpits of those airplanes. And yet the terrorists accomplished their destruction with the aid of both the planes and the buildings that swallowed them. The towers stood and collapsed as a result of the complicated interplay between the designers' efforts and those of the terrorists.

Communication and Conclusions

If the early stages of engineering problem-solving involve isolating and narrowing the problem into engineering terms, this final stage requires integration. It is here in the process of recombination that we can reassess what we know, what we've accomplished, and what gaps remain. The very act of writing or speaking helps to clarify our understanding and is thus as essential as any of the earlier stages of engineering problem-solving. As the stymied rescue efforts in the

Twin Towers make clear, having knowledge but not the ability to communicate it can have devastating consequences.

The collapse of the Twin Towers was not a typical engineering event, yet it illustrates some commonplace characteristics of the engineering approach to problem solving. The task of reconstructing what had happened had begun to be shaped before the engineers ever set foot in that Fresh Kills salvage yard, by the happenstance of what evidence survived the collapse and by the decision to place speed of cleanup ahead of preservation of evidence. The engineers who arrived shaped their task as well, both by what they chose to examine and by what they chose to ignore. Although the goal may have been an objective analysis of "building performance," the result was constrained and shaped by what was made available to the engineers and by what they brought to the project.

The FEMA/ASCE engineers recognized many of the external limitations of their study—the piecemeal evidence, the shortage of funds. Engineering has, after all, been defined as design under constraint. Whether these same engineers thought of themselves as contributing to those constraints is less clear. The point is not to blame these engineers, who, like Leslie Robertson, did an admirable and difficult job, but to suggest that one lesson to be learned is that engineering cannot be as detached as we may be inclined to think.

As was true for so many Americans in the weeks after 9/11, engineers sought to contribute in what ways they could, namely, by offering up their technical expertise to the efforts at Ground Zero. The disputes at Ground Zero over responsibility for the work to be done there led to numerous turf battles. Engineers rightly focused their efforts on their unique skills and knowledge. The FEMA/ASCE study team had no power to turn back the clock and no authority to assign responsibility, but they possessed the ability to identify ways to strengthen existing buildings against future disasters. The FEMA/ASCE report allows that the structure itself may have been at fault—indeed, by limiting their focus to the performance of the buildings, the study team could hardly have come to a different conclusion than that the buildings were involved in their own collapse.

William Langewiesche described the post-9/11 engineering project at Ground Zero as one of "unbuilding," and it is one that engineers tackled with remarkable talents and expertise. However, if we attempt a metaphorical unbuilding of the towers from only the engineering standpoint, the result will be very shaky indeed. Surely the terrorists are primarily to blame for the devastation of September 11, whatever the engineers did or did not do, and this simple claim should be sufficient to remind us that engineers and their designs do not stand (or fall) alone. The success of any structure depends upon wise decisions being made throughout the design process by both engineers and nonengineers, from initial choices about whether to build at all, through complex judgments about the tradeoffs inherent in design, to how the final structure is viewed and used.

In short, the horrors and the heroism of September 11 had complex roots that we should not ignore as we explore the mechanism of the collapse. Surely, there is benefit to focusing on technical questions regarding the sequence of the collapses. But the lessons to be learned extend outside the confines of a technological black box. The goal of the "painful probing" that the *New York Times* encouraged officials to pursue should not be to determine whether the towers and their designers were *either* successes *or* failures, but to articulate the ways in which they were *both* successes *and* failures. To grasp such complexities is not a sign of indecisiveness, but rather an indication of sophisticated understanding that characterizes a professional. Leslie Robertson described being an engineer as "a tremendous responsibility" because of the ambiguity of engineering practice: "It's a very imperfect process. It's not so beautiful as science."[45]

FOR FURTHER EXPLORATION

1. Locate an article or video about the WTC collapse, another disaster, or another engineering project. Which of the five steps of design (or investi-

gation) are covered and how? What are the pros and cons to the approach and the message in the source you selected? How might the description change if it included more (or fewer) of the five steps?

2. Describe the ways in which an undergraduate engineering curriculum (including both engineering and nonengineering courses) fits in with the stages of engineering problem-solving. Which courses/subjects will help develop skills in each of the five areas? Are some areas likely to be over- or underdeveloped? In what ways does the curriculum encourage or discourage a broad perspective on engineering design?
3. The U.S. engineering accreditation agency ABET requires that graduates of accredited programs have "a knowledge of contemporary issues." What sort of contemporary issues are relevant and how? How might one gain this knowledge? Select any current event (e.g., in today's newspaper)—is it significant to engineering practice? Are any topics fully outside of engineering? If so, where should the line between relevant and irrelevant be drawn?

CHAPTER TWO

"Finding Hope in the Ruins"

A Short History of Engineering Disasters

We want to make it right somehow.

HASSAN ASTANEH

In part, what the engineers responding to 9/11 were wrestling with was not only how to respond to the collapse of the Twin Towers, but how to make sense of the collapse at all. What did the devastation at Ground Zero say about the engineering profession? What role does failure have in expertise? It is one thing to say that we should learn from past failures. It is another to suggest that failure should never occur.

Here we find another of the contradictions in engineering practice. We want to minimize failure and learn from it, and yet on some level we have to expect that it will happen. We have to begin and end every project by admitting that engineers are human, as are the users of our creations. Mistakes will happen. Indeed, some would argue that failures are the most valuable learning tools in engineering and should be embraced, if not welcomed.

As one journalist described it, engineers, like the public at large, tried "finding hope in the ruins" as one method of coming to terms with the disaster. One engineer explained that the investigations offered a form of redemption: "We want to make it right somehow. Make the next buildings stronger." But to be honest, we're not very good at reconciling success and failure in engineering. To begin to make sense of this contradiction and to understand more deeply the challenges faced by the WTC investigators, let's step back and take a look at how American engineers have talked about past failures and the role of engineers in responding to them.[1]

Failure as Lack of Expertise: Defining a Professional Engineer

In May 1873, a 660-foot-long wagon and foot bridge across the Rock River at Dixon, Illinois, fell into the river "with a quick crash," killing 45 of the 150 spectators who had been observing a baptismal service being held in the waters below.[2]

If the Dixon accident had been unique, it might, according to many civil engineers, have been "considered inevitable and unforeseen." But the real tragedy was that the Dixon bridge collapse was only the latest in a series of bridge failures. Another bridge of the same patented Truesdell design had fallen in December 1868, been repaired and reinforced, and had fallen again in July 1869. Some historians have estimated that as many as one in four nineteenth-century bridges failed. Though these figures have been disputed, nineteenth-century Americans clearly believed that the rate of failures, whatever it may have been, was unacceptably high.[3]

Decrying "the frequent marine disasters, boiler explosions, and kindred horrors that have crowded upon us of late," *Scientific American* declared that "any competent engineer should have been able to perceive at a glance" that the Dixon bridge was "improperly built and unsafe." But the editors supposed that nothing much would be done, saying "it seems an almost useless task" even to call for an investigation. The local city council had ignored earlier suspicions about the bridge's strength; the "bereaved," though they were "indignant" and calling "loudly for the exposure and punishment of the guilty parties," would soon turn instead to healing their wounds and moving on; the general public would only be "shocked by the sensation for a day or two."[4]

The directors of the American Society of Civil Engineers also expressed concern over this string of tragedies. As they began to consider what the society's role should be in dealing with failures, they focused less on the deaths themselves (about which they could do nothing) and more on the impact of those deaths on the public image of engineering (about which they felt great apprehension) and how to prevent similar collapses in the future. The directors responded to this "calamitous disaster . . . and other casualties of a similar character" by commissioning a report, "On the Means of Averting Bridge Accidents."[5]

The commission comprised seven of the most prominent bridge engineers of the day, including James Eads, who served as chairman, and ASCE president and founding member Julius Walker Adams. After determining that sources of the "disastrous accidents" of the past few years fell into three categories—building, maintenance, and use—the committee suggested means of preventing each type of failure. The first class of problems included bridges "erected by incompetent or corrupt builders." The second class consisted of "good designs" ruined by neglectful owners or careless transportation of the bridge's materials. Class three included bridge accidents brought about by unsafe use, in the form of unanticipated speed or amount of traffic, for instance. The committee members proposed that all three types of failures could be reduced or prevented by establishing a formal set of standards for bridge building that would entail minimum specifications for design, mandatory inspection during construction, and periodic reviews to evaluate any unexpected changes in the use or maintenance of the structure.[6]

But the committee members could not agree on two points: who should propose and set the standards, and should they be mandatory or voluntary? At issue was how much of a role the society should play in the oversight of civil engineering. On the one hand, the society's membership included some of the most qualified engineers practicing in the United States. These members of the ASCE—"the most competent authority in the premises"[7]—were not likely to be personally responsible for failures, according to the Dixon report, but perhaps they had an obligation (as members of the society or as members of the profession) to assist other engineers by sharing their knowledge and expertise. On the other hand, adopting a set of standards directed solely at engineers (the only group the ASCE could reasonably address) would imply that engineers rather than owners were generally responsible for bridge collapses—a misimpression the committee members clearly wanted to avoid.

To "assert itself as the only competent authority on bridge construction" was, in the end, more of a burden than most society members were willing to accept. They argued that merely establishing a set of specifications could instill too much confidence in unqualified builders, who, in the words of one society member, could not "be prevented from building bad bridges by any specifications, however elaborate."[8]

The findings and recommendations of the Dixon committee grew out of the nature of the society in those years. The ASCE was in the nineteenth century a small and selective society dedicated to providing fellowship and status to those few who were invited to join. In an era before college education became common for American engineers (and when licensure was still several generations away), membership in the society served as a diploma of sorts, certifying the qualifications of its members. Because nineteenth-century American civil engineers relied so heavily on their reputation rather than formal documentation of their abilities, any engineering failure posed a threat to the entire profession by raising questions about those abilities. As the leading authorities in civil engineering, ASCE members deemed themselves well suited to establishing the causes of failures and felt obliged not only to investigate the causes of such failures, but also to separate themselves from those responsible for the failures. Society members took for granted that such causes were to be found outside the ASCE.

The attention the society gave to the failure of the Dixon bridge paled in comparison to their concern over the collapse of the Ashtabula bridge three years later. The Ashtabula stirred more discussion in the ASCE than any other nineteenth-century failure. It was dramatic, deadly, and problematic.

On a snowy night in late December 1876, two locomotives in tandem started across a wrought-iron truss railroad bridge over Ashtabula Creek in Ohio, pulling eleven passenger and freight cars behind them. The engineer of the first locomotive, Dan McGuire, felt a sudden drag as his engine started across the bridge. He opened his throttle wide to escape the trouble, whatever it was, only to find that his locomotive was "sideswiping the bridge abutment." McGuire then heard "a terrific crash" as the second locomotive slammed head-on into the abutment. The crash uncoupled the two engines; McGuire sped off the far end of the bridge in his, while the second fell with the bridge into the chasm below, dragging all eleven cars behind it. The fall and the ensuing fire killed 80 of the 134 passengers and crew.[9]

One early explanation offered for the collapse was that some of the front wheels or trucks of the westbound train had become derailed in the snow east of the bridge. The derailed wheels then tore up the floor of the bridge, weakening the structure and initiating the collapse. The coroner's inquiry, however, found that some of the bridge braces had been coming loose before the train arrived, indicating that defects in the bridge itself had caused the derailment, which in turn led to the failure.[10]

The press attacked the bridge designer, an engineer named Amasa Stone, as having attempted "a difficult piece of construction, with but little specialized knowledge of the principles involved in the task." The ASCE members who investigated the bridge collapse decided early that Stone could not have been at fault for the failure. Committee member Charles MacDonald, for example, countered that "the Ashtabula bridge was the result of an honest effort to improve the bridge practice of the country, undertaken by a man whose experience in wooden bridges warranted him in making the attempt." Stone had used the common and previously reliable Howe truss style, invented by his brother-in-law, on the bridge.[11] Stone's design specified that the bridge be constructed entirely of wrought iron, and the Ashtabula site required the span be longer than any previous Howe truss bridge—two significant departures from the trusted method. What the media portrayed as foolhardy, the ASCE investigators called daring. Describing the Ashtabula bridge as "an exceptional structure," Charles MacDonald argued that its failure should not impair "the reputation of American engineers and bridge constructors."[12]

MacDonald and others instead directed the blame at the bridge inspector. Charles Collins, the railroad's chief engineer, but not an ASCE member, was a "sensitive, gentle man," who, according to both the coroner's inquest and the Ohio legislative committee investigation, had examined the bridge "frequently and conscientiously." Although the coroner and the legislative committee had both found Collins not at fault, several members of the ASCE argued that "the inspection must have been faulty," since "any bridge expert would have condemned the bridge almost on sight." One ASCE member declared confidently that if all bridges in America were inspected by a qualified bridge engineer, "in less than two years there need not be a dangerous iron bridge in America, or a bad one built hereafter." The trouble in Ohio, according to several ASCE commentators, was not the lack of inspection, but the poor qualifications of the inspector. One ASCE investigator hoped that the Ashtabula failure would convince "railroad managers" like Charles Collins "not to attempt too much engineering themselves." In spite of his exoneration by two investigative committees and the fact that Collins "had had nothing to do with the design of the bridge," this sensitive, gentle man resigned his post and then killed himself.[13]

ASCE members expressed concern not only about the bridge collapse itself but also about a number of legislative proposals made "under the spur of the panic derived from the Ashtabula disaster." James Garfield, then a U.S. representative, had introduced a bill "to provide for a more thorough investigation of accidents upon railroads." Garfield's bill authorized the president to appoint a committee of three army engineers to study recent failures and to investigate future accidents "of an unusual or unexplained character." In addition to this federal effort, several states began to pass laws similar to Ohio's new Bill to Secure Greater Safety for Public Travel over Bridges. These state laws established minimum loads for bridge designs and required that all bridges be inspected "by some competent and suitable person" at regular intervals (determined by the length of the span). ASCE members, including past-director Thomas C. Clarke, objected that since neither the federal nor the state laws required civil engineers to be involved in the design, construction, or inspection of bridges, "the enforcement of these laws will be placed in the hands of men of very varying qualities."[14]

Society officer Charles Shaler Smith steadfastly declared that had the standards he and others proposed in the Dixon report been supported by ASCE members in 1875, not only would these new statutes be unnecessary, but "the Ashtabula bridge would have been condemned and made safe, and the horrors of that fatal night averted." In the end, Smith's proposal for society-approved recommendations did not prevail. Society members understood the problems bridge inspection statutes and standards were meant to address, but felt that such oversight of the profession was not a duty of the society or its members, and they did not wish it to be.[15]

American civil engineers did not deny that failures occurred in these years. Disasters like the Ashtabula were well known to engineers and the public alike. But failures, according to the ASCE, were the result of someone else's incompetence. The only role for eminent engineers (such as ASCE members) to play in failures was to investigate and prevent them. Their confidence was perhaps not entirely misplaced. As bridge historian Joseph Gies has noted, triumph accompanied disaster in these years. In 1874, the Eads Bridge, named for its chief engineer, opened in St. Louis. Roebling's Brooklyn Bridge was completed in 1883, the same year the first skyscraper graced the Chicago skyline. According to Gies, such feats "pulled . . . engineering all the way from the eighteenth to the twentieth century." But in that same span of time, failures like the Ashtabula, along with uncounted minor failures, "had thrown the engineering profession into a state of confusion, not to mention their effect on the public." Though they may have been suffering from confusion, engineers remained sufficiently clear-headed to emphasize their triumphs over their defeats.[16]

These successes had convinced civil engineers that good engineering resulted from properly applying known scientific principles. In the words of one of the most famous of the nineteenth-century bridge builders, James Eads, "The laws which guide an engineer are immutable, and never deceive. Failures and disasters . . . result almost invariably from the non-observance of these laws, or from a want of knowledge of them." According to historian Sara Ruth Watson, "Eads was confident that Nature was not only knowable but known and that technical knowledge and human will could master the universe. Truth was absolute, engineering an exact science."[17] As we will see, such certainty in the power of engineering science has not evaporated, but for our present purposes, it is important to note how powerfully and thoroughly this view shaped the ASCE's response to failures in the nineteenth century.

This equation of success with professionalism and failure with lack of expertise was tested after the Ashtabula bridge collapse but remained the predominant view for the rest of the century. By 1907, however, the president of the ASCE, George Benzenberg, would concede that "the engineer is not infallible; he may make mistakes, and it would be remarkable indeed if he did not at some time err in his judgment." However, he argued, providing the engineer "is conscientious, exercises every care, and exhausts all his resources in the performance of his obligations, it cannot be said that he is culpable or negligent in his duty."[18] In other words, according to President Benzenberg, when engineers make mistakes they should not be held responsible. A number of major projects around the turn of the century helped support this new conception of success, failure, and responsibility, but one engineering project in particular ensured that the earlier view of failure as solely the result of inexpert practice was no longer tenable.

Failure as the Price of Progress: Authority on the Quebec Bridge

Building a bridge across the St. Lawrence River at Quebec was a big project—some would say unprecedented. Such a monumental task required the skills of a monumental engineer. Theodore Cooper was just such a prominent engineer, and furthermore, he wanted the job. To early-twentieth-century engineers and the public alike, it was inconceivable that so huge a project, headed by so talented a man and watched over by so many, could fail. But it did. Twice.[19]

Since the ASCE represented only the most elite engineers, a failure involving a society leader called into question all that the society members believed in—the very basis of their claims to professionalism. Engineers, after all, were building America. Monuments like the Brooklyn Bridge and the new skyscrapers punctuating the urban landscape were the work of leading engineers. Failures such as the Dixon and Ashtabula bridges were the work of lesser men and, if anything, served to reinforce the importance of hiring only qualified engineers. And so, in late August 1907, it was with great shock that American engineers awoke to news of the Quebec Bridge disaster.

The "Quebec Bridge" was the only name by which the structure had been known. Located five miles downstream of Quebec, it was in mid-construction that August, still two years from its scheduled completion, thanks to delays resulting largely from lack of funds. The two anchor arms of the bridge from each shore to the mid-river supports had been completed, and work had begun on the center span by cantilevering out from either buttress. The two halves were due to meet in the middle in 1909. On August 29, 1907, as workers waited for the whistle to blow ending their day's work, they sent the traveler—a small train car used for moving heavy equipment and materials along the bridge—out the arm on the south (Quebec) half of the structure to deliver materials for the next day's work. As the traveler neared the end of the arm, the whole Quebec side of the bridge buckled with "a loud report like a cannon shot," and the mighty capstone to Theodore Cooper's career crashed into the St. Lawrence River below, sending up a rumble "plainly heard in Quebec." Cries of workers trapped amid the nineteen hundred tons of steel continued into the night. Only 11 of the 86 workers on the bridge survived.[20]

By the late summer of 1907 the bridge, stretching majestically over the river, had already been an impressive sight, and engineers were marveling at the accomplishment in progress. *Engineering News* and other periodicals had reported on the status of the bridge construction from the start, knowing the completed bridge would make history as the longest-spanning arch in the world. The growing span, all were agreed, was a testament to the skills and experience of Theodore Cooper.[21]

Born in 1839, Theodore Cooper graduated from Rensselaer in 1858 with a degree in civil engineering. His diploma alone made him one of the best-educated engineers in the country. But Cooper had also trained with the best, working under James Eads at Midvale Steel and assisting Washington Roebling on the Brooklyn Bridge. He became a member of the ASCE in 1874 and served as a society director just 10 years later, at age 45. Cooper frequently contributed articles and commentary on the science and art of bridge building to the pages of ASCE publications and *Engineering News*. Several of these contributions, including a statement on the Ashtabula Bridge, covered engineering failures. He had won the society's Norman Medal for two of his essays on bridge construction. He was, according to a colleague, "universally familiar to bridge engineers," which served to make his role in the Quebec failure all the more troubling.[22]

From the earliest days of the Quebec project, Cooper had sought out and defended his authority over the project. At one point, the Canadian Department of Railways and Canals' chief engineer suggested that the department hire its own engineer to review the plans once Cooper had done so. Cooper angrily replied, "This puts me in the position of a subordinate, which I cannot accept." By the time the negotiations were complete, "everything of import was referred to Cooper."[23]

Unfortunately, there emerged at least one serious drawback to Cooper's being in such complete control. Before construction on the bridge even began, Cooper realized that his declining health would prevent him from traveling to Quebec. He offered to resign his post but acknowledged that his withdrawal from the project would further delay the already long and expensive project. The project organizers agreed to keep him on, and Cooper "took no further action" to have himself replaced. Poor health did indeed keep him in New York, forcing him to leave inspections to his on-site assistant, N. R. McLure, who reported to Cooper weekly with updates on the progress of construction. Cooper visited the site only three times and never once saw the actual bridge during construction.[24]

The front page of the August 30 *New York Times* carried the horrid news: "Bridge Falls, Drowning 80." (It would later become clear that 75 had in fact died.) That such a structure could collapse was hard enough to believe, but the next day Theodore Cooper's name appeared in the *Times'* front-page headline. Cooper revealed that the very day of the collapse he had sent a telegram to Quebec ordering all work on the bridge to stop until further notice, following a report from his inspector. He told reporters, "I feel a little guilty about this. I suppose commissions will investigate and locate the responsibility."[25]

Theodore Cooper had been having doubts about the bridge project since at least early August. On August 7, his assistant McLure sent word to New York that some of the rib chords arriving from the fabricators in Phoenixville, Pennsylvania, were bent. On the afternoon of August 29, having received yet another report from McLure, this time in person, Cooper allowed his fears about the unseen project to stir him to action. Though he believed he had authority over the design of the bridge, he did not think he was in charge of the construction, and doubted "whether a foreman [on site] would take a suggestion from me or not." Nonetheless, Cooper felt it was his "duty to warn the building company to suspend work until the bridge could be examined." As a result, he telegraphed, not the on-site supervisor, but his own superiors at the Phoenix Bridge Company in Pennsylvania, intending for them to wire the foreman at Quebec to stop the work on the bridge. Cooper's wire contained only the brief message: "Add no more load to bridge till after due consideration of facts. McLure will be over at five (5) o'clock." Cooper believed that there was a serious problem with the bridge but did not think trouble "was so imminent that a remedy could not be applied." He sent McLure to Phoenixville to reinforce the telegram and to discuss the problem with the Phoenix Bridge Company. McLure arrived in Pennsylvania at 5:15 to explain the severity of the potential trouble, but too late. By 5:30, 75 men in Quebec were dead.[26]

The Canadian federal government appointed a commission staffed by two practicing Canadian civil engineers and a Canadian professor of engineering to investigate the cause of the bridge collapse and report the result of its inquiry, "together with . . . any opinion they may see fit to express thereon." In November 1907 the Canadian commission questioned officers and engineers from both bridge companies, a Quebec Bridge Company inspector working in Phoenixville, and most of the survivors of the collapse. The testimony included "vigorous language directed against Theodore Cooper."[27]

The Royal Commission of Inquiry presented its report in 1908, placing "the brunt of the blame" on Theodore Cooper and designer Peter Szlapka, saying "the failure cannot be attributed to any cause other than errors in judgment on the part of these two engineers." The commission found that Cooper had exercised inadequate oversight of the project and had paid insufficient heed to workers' warnings of danger on the bridge. The commissioners went on to say that "these errors of judgment cannot be attributed either to lack of common professional knowledge, to neglect of duty, or to a desire to economize. The ability of the two engineers was tried in one of the most difficult professional problems of the day and proved to be insufficient for the task."[28]

Bridge collapses were not uncommon in these years, and engineers were accustomed to news of them. But two aspects of the Quebec Bridge collapse made it especially remarkable. First, the Quebec Bridge was a major and daring project; thus, engineers (and the public) had great confidence in its success. Earlier large-scale projects such as the Eads Bridge in St. Louis and the Brooklyn Bridge in New York had been completed successfully. The few deaths during construction of such projects, though unfortunate, were seen by engineers and by much of the public as the price to be paid for such grandeur. The very newness of these projects meant extra care was put into designing and building them. Second, such projects were overseen by the best engineers in the business. Eads, Roebling, and Cooper knew more about bridge building than anyone else. As the ASCE had been arguing for years, the work of such seasoned engineers ought to have been more reliable than that of a young, inexperienced, or untrained engineer. Given Cooper's prominent position in the ASCE, there was probably nothing the society could have said that would have helped the society members or the public understand how the collapse could have happened. For the first time in the society's 50-year existence, the ASCE opted not to convene a study or issue any report on the failure of a civil engineering work.

Tragically, the Quebec Bridge would fall again during its rebuilding in 1916, and would not be successfully completed until 1917.[29] As the central span was being lifted into place in 1916, it slipped from its supports and crashed into the river, killing 11 workers. In an astonishing article, *Engineering Record* noted the profession's desire to "clear up the mystery" of this second collapse quickly since the installation technique was

> a great step forward in bridge-erection practice. It was originated and developed by the best bridge-engineering talent on the American continent. Its details were thought out with a minuteness and care that were nothing short of marvelous. Moreover, it worked—yes, we repeat it—it worked even tho the

> span lies to-day in the bed of the river. . . . The loss of the span, while it may in the lay mind cast discredit on those responsible for the work, in reality put the remaining parts of the structure to a most extraordinary test, and so proved the ability of the designers and builders.[30]

Again the profession steadfastly refused to allow that a true and talented engineer could be responsible for a disaster involving an engineering project. Other failures, equally dramatic and equally puzzling, would continue to trouble the profession throughout the twentieth century. A generation after the Quebec Bridge collapses, the ASCE would once again wrestle with the role of otherwise competent engineers in the failure of their structures.

The Tacoma Narrows Bridge, known as Galloping Gertie for its tendency to sway in strong winds, twisted itself apart on the morning of November 7, 1940, four months after completion. The bridge had given sufficient warning to prevent anyone other than a reporter's dog from dying in the collapse, but the unusual behavior of the bridge attracted a great deal of attention in the engineering community.

When the bridge fell, engineers realized that they had less understanding of bridge aerodynamics than they had believed, and in this sense, the Tacoma failure parallels the Quebec Bridge disaster over 30 years earlier. Since engineers had declared "Galloping Gertie" safe despite her strange bouncing, they were hard pressed to admit they had been mistaken. Instead, they focused on pointing out that, given the state of knowledge of harmonic oscillation, they could not have predicted that the bridge would fail, just as an earlier generation of engineers had praised Theodore Cooper's work on the Quebec Bridge, saying he had no way of knowing that structure was as unsafe as it turned out to be.[31]

A retrospective article published by the ASCE in 1965 on the lessons that had been learned in the quarter-century since the Tacoma Narrows failure mentioned the bridge's designer, Leon S. Moisseiff, by name, calling him "a distinguished bridge designer . . . who had contributed a significant slice of knowledge to the engineering profession." Like Theodore Cooper before him, he was an ASCE member who, although his unfortunate structure had failed, had been pushing the limits of engineering knowledge. Moisseiff was not accountable for the failure because, unlike an incompetent designer, he knew more, not less, about bridge engineering than most engineers. Here, a half-century after the Quebec Bridge created a crisis of faith among engineers about their fallibility, lay the answer to how a competent, knowledgeable engineer could play a role in such a devastating event. According to the 1965 ASCE article, "it was ignorance on the part of the entire engineering profession, not solely on the part of Moisseiff, that led to 'Galloping Gertie's' ruin."[32]

By the mid-twentieth century, then, American civil engineers had identified two key types of failure—those that resulted from unqualified practitioners pushing the bounds of their own knowledge and those that resulted from well-qualified practitioners pushing the bounds of the profession's knowledge. In 1981, the profession was faced with a disaster resulting from a mistake so simple that it could be readily identified by a sophomore engineering student and yet had apparently been made or missed by a pair of unquestionably well-qualified engineers. This devastating event challenged everything American engineers had taken as givens in their understanding of professional responsibility and would require that they begin to consider failure not as a preventable occurrence with no place in engineering, but as an unavoidable event inherent in their endeavors. The former conception viewed failure as something that could and should be eradicated; the latter viewed failure as something to be managed and ultimately accepted.

Failure as Inherent in Systems: "The Hyatt Horror"

July 17, 1981, was a humid Friday evening in Kansas City, Missouri. Between 1,500 and 2,000 area residents chose to escape the heat at the Hyatt Regency Hotel's tea dance, a weekly event featuring big band music and a dance contest. Some came to celebrate birthdays or anniversaries; others, to end another work week with a relaxing night out amid the unusual architecture at the Hyatt Regency. Delicate steel rods extended from the ceiling of the airy five-story atrium; on these rods hung three graceful walkways that spanned the lobby at the second-, third-, and fourth-floor levels.[33]

Several groups of friends gathered on the walkways to watch the crowd of partygoers below. Just after seven o'clock, a thunder-like crack shook the lobby as the fourth-floor walkway tore loose from its supports and dropped 30 feet to the lobby floor, crushing the second-floor walk beneath it. Guests on the offset and thus still-suspended third-floor walkway watched in dismay. Hundreds of guests lay trapped under the more than 70 tons of debris. The final count of 114 dead and nearly 200 injured led one group of investigators to declare the Hyatt disaster "the most devastating structural collapse" in U.S. history. The local newspaper called it simply "The Hyatt Horror."[34]

Daniel M. Duncan and Jack D. Gillum, two key engineers on the Hyatt design team, later expressed "shock" and "dismay"—not only at the nearly 300 dead and injured, but also at the severity of their punishment. Their licenses had been revoked by the Missouri licensing board, making them the first American engineers to lose their licenses for gross negligence.[35] After more than a century of equating engineers with success, the civil engineering profession had been forced to suggest that a good engineer could practice engineering badly. It was a disquieting admission, and one that left many in the profession ill at ease.

One of the first investigations into the Hyatt collapse came from the National Bureau of Standards. What took the NBS ten months to investigate and document, the local newspapers had discovered in four days, and engineer Jack Gillum himself recognized on sight. Investigators traced the walkway collapse to poor connections between the walkways and the hanger rods from which they

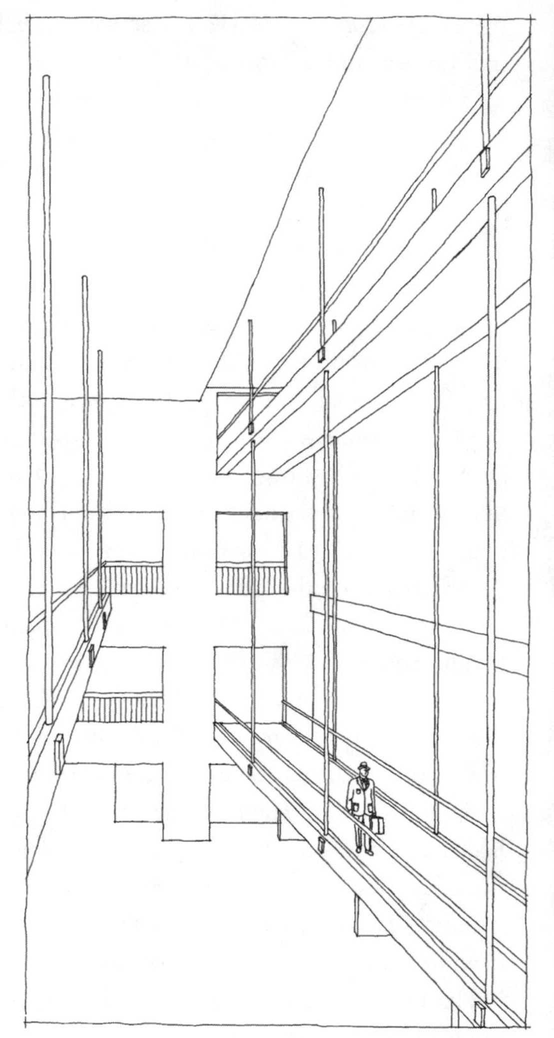

Hyatt Skywalks Schematic. Layout of the three skywalks crossing the lobby of the Kansas City Hyatt Regency. The walkways at floors two and four appear on the right-hand side, one under the other. The third-floor walkway, which did not collapse, appears on the left. NBS, *Investigation of the Kansas City Hyatt Regency Walkways Collapse*, www.fire.nist.gov/bfrlpubs/build82/PDF/b82002.pdf, p. 21.

were suspended. The design of the skywalks had called for both the second- and fourth-floor walks to hang from a single set of rods suspended from the lobby ceiling and running through both walkways. (Picture two people each hanging off a rope attached to the ceiling.) But the skywalks were not built according to the original plan. Instead, the fourth-floor walk hung from rods connected to the ceiling, and the second-floor walk hung from an additional set of rods connected to the fourth-floor walk. The change effectively doubled the load on the fourth-floor walkway connection, which was clearly not up to the task. (Imagine the lower person now hanging from the upper person, who now must work twice as hard to keep a grip on the rope.) An obvious question loomed: who could have made such a simple and deadly mistake?[36]

Much like the *Building Performance Study* for the World Trade Center, the 1982 NBS report made it clear that its purpose was not to assign responsibility for the collapse and therefore attributed the

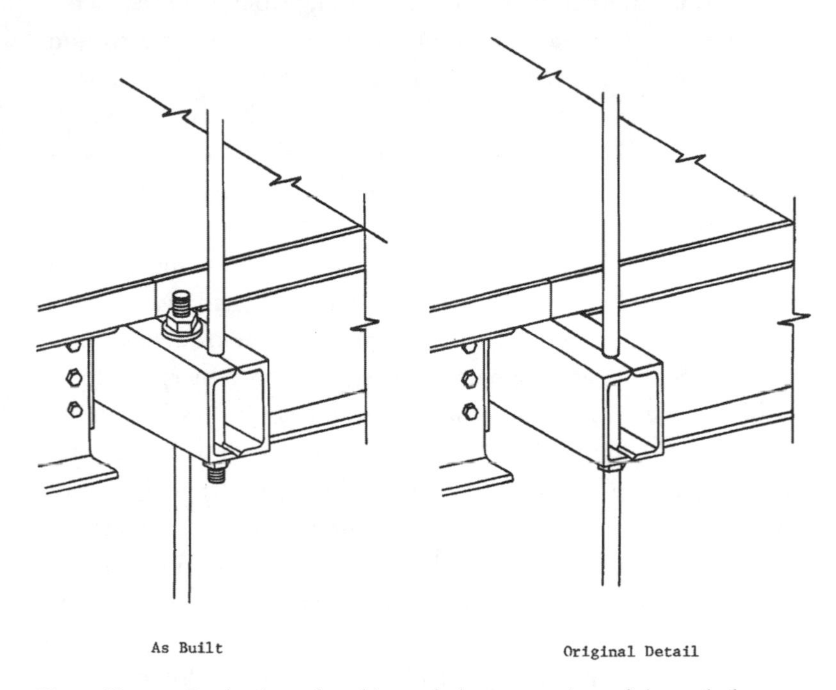

Hyatt Hanger Rods. Actual and intended construction of the rods from which the walkways hung in the Kansas City Hyatt Regency. The change, resulting from a series of miscommunications, doubled the load on the connection between the upper rods and the fourth-floor walkway and led to the collapse that killed 114 people. NBS, *Investigation of the Kansas City Hyatt Regency Walkways Collapse*, http://www.fire.nist.gov/bfrlpubs/build82/PDF/b82002.pdf, p. 251.

Hyatt failure to design error without ever asking about the role of the designers. Upon receiving the NBS report, the Missouri licensing board began quietly to pursue its own inquiry, but did not want to be seen as the sole source of justice in the case. As a result, the board, according to chairman Paul Munger, "took steps to shelter itself. It held no press conferences and did not present any fanfare over its work."[37]

Though the board kept its actions quiet, the public was well aware of other aspects of the tragedy. The local newspapers repeatedly published gruesome, minute-by-minute accounts of the collapse and its aftermath, covering everything from what the victims had eaten for breakfast that morning to how many severed body parts had arrived at the morgue that night. Such detailed descriptions haunted even those who had not been at the Hyatt or known someone there. Hardly surprising, then, that the licensing board did not wish to call attention to the possible role of engineers in such devastation.[38]

Despite the licensing board's secrecy in examining the case, the board members could not ignore what they had discovered in their investigation. As simple as it was to see the trouble with the fourth-floor walkway connection, what the board members found in the organization of the project was almost hopelessly complex. The chain of command for the Hyatt construction was staggering, involving multiple corporations and subsidiaries stretching across several states, with an intricate system of coalitions, contracts, and subcontracts for the design and construction of the hotel and the redevelopment project of which it was a part. The architects hired Jack D. Gillum and Associates, a prominent St. Louis engineering firm that had worked on earlier phases of the redevelopment project. Jack Gillum, as owner of JDGA, served as supervising engineer on the Hyatt, and he assigned Daniel Duncan to be the project engineer, in charge of day-to-day work. The project also required, among others, fabricators, who worked both as members of the design team, handling simple design details as allowed by law, and as members of the construction team, obtaining the materials the construction team needed. This complex hierarchy left the licensing board with an equally complex task of assigning responsibility for the faulty walkway connections.[39]

A further complication added to the confusion on the Hyatt; the hotel was a fast-track construction project, meaning that the construction team had begun to build the hotel while the design team was still finalizing the plans. In the midst of this complexity, the design of the connection between the skywalks and their supporting rods became muddled. The engineers left the detail unspecified, indicating that the fabricators were to complete the calculations for the design. The fabricators later argued that the connection was not their responsibility. The engineers' sketch seemed to indicate that a single set of hanger rods should connect both walkways to the ceiling, yet somehow the walkways were built with two sets of rods. The engineering firm's records show that Jack Gillum's seal appears on the plans, indicating that the engineering firm agreed to the design, no matter who had suggested or calculated it.[40]

If Duncan and Gillum had merely overlooked the dangers of the new design, the accident could perhaps have been passed off as a very costly yet innocent mistake, but the faulty connection had been the subject of concern before the accident. The board's investigation revealed that the project engineer, Daniel Duncan, had been asked about the implications of the design change on at least six separate occasions during construction. Duncan seemingly assured each inquirer that replacing the single, long hanger rods with double, offset rods would not compromise the safety of the walkways. Duncan later testified that the connection and any changes to it were not his responsibility because the engineers had not designed it in the first place. As a result, when he had been asked about design changes to the rods, he intended his answer to indicate that his firm had no concerns about someone else doing a redesign. It was a simple, if repeated, miscommunication with deadly consequences.[41]

The licensing board members faced a difficult decision. Since its establishment more than 40 years earlier, the Missouri board had declared that its mission was to protect the public, not to punish engineers. Its operating policy had been to "rehabilitate when at all possible" and to avoid disciplinary action.[42]

Though Duncan and Gillum were forced during testimony to admit they had nominally approved of the change (and had subsequently not caught the error), they refused to concede that this made them responsible, particularly since there was no proof as to who had altered the plans. The Missouri licensing law stated that a professional engineer is "responsible for the contents of all" documents on which he places his seal. Jack Gillum and Daniel Duncan, however, claimed that because of the complexity of most engineering projects and the number of people involved in the design process, many professional engineers routinely sealed plans that they had not personally checked. Judge Deutsch, who was hearing the licensing board case, pointed out that even if such a practice of sign-offs was common, it was still "unreasonably dangerous." He further noted that "while the engineer may properly delegate the work of performing engineering design functions, he cannot delegate the responsibility."[43]

On November 15, 1985, Judge Deutsch filed his decision. He found both Duncan and Gillum guilty of gross negligence and misconduct and added unprofessional conduct to Gillum's list of transgressions. Deutsch gave the licensing board permission to revoke the engineers' licenses. Despite receiving more than 60 letters of protest, mainly from Duncan and Gillum's friends and colleagues, the board carried out the punishment on January 22, 1986, four and a half years after the walkway collapse.[44]

Meanwhile, the ASCE struggled to make its own decision as to how to pursue the case. Throughout the NBS and licensing board investigations, the society had continued to view its mission as promoting the profession, not policing it. Though the code of ethics had been revised in 1976 to make public safety "paramount," enforcement of the code had never become a priority. Daniel Duncan was free from any discipline by an ethics committee because he had never been a member of any national society. The ASCE had still not taken any public action against Jack Gillum.[45]

But now, faced with the fact that the licensing board had gone so far as to revoke the two licenses, the ASCE Committee on Professional Conduct felt obliged to address the case. The committee responded with a confidential hearing on the matter during the summer of 1986. Largely on the basis of the licensing board's evidence, the committee members concluded unanimously that Gillum should be "expelled with no privilege ever to rejoin" and submitted their recommendation to the board of direction. After more than 12 hours of heated deliberation, the board members chose to ignore that recommendation and instead found Gillum "vicariously responsible . . . but not guilty of gross negligence nor of unprofessional conduct." They voted to suspend him for just three years. Gillum voluntarily relinquished his membership altogether. Though the board's decision appeared in ASCE publications, society policy kept additional details confidential.[46]

The members of the licensing board and the ASCE directors had clearly agonized about how to respond to public calls for accountability while acknowledging the realities of engineering practice. They had been reluctant to begin their investigations, made every effort to conduct them quietly, and engaged in lengthy debate about the right determination. Letters and articles in the technical press during these years indicate that a great many members of the profession were equally concerned about how to resolve the question of who was responsible for the walkways' collapse.[47]

In essence, the struggle following the Hyatt disaster resulted from a conflict between a desire to identify with certainty the root cause of the collapse and a conviction that Duncan and Gillum's work—in so many ways the model of good engineering—could not possibly have been to blame for such devastation. The work of sociologist Charles Perrow provides an explanation for how good work can lead to bad outcomes. Perrow argued in his groundbreaking work *Normal Accidents* that complex systems have outstripped our ability to perfect them. In any design with millions of details (and dozens of designers), it becomes impossible to predict with certainty how the whole will behave. Engineers have long understood that connection points are the areas of design most prone to failure: it is simply easier to predict and control how a single component will behave than how two components will interact. In complex systems, the interactions become too numerous and complicated to control, and, even more frustrating, any attempt to perfect one part of the system will result in unforeseeable consequences for other parts of the system.[48]

In short, Perrow argues that complex systems are not perfectible and that failure is an inherent and necessary part of them. Perrow cautions us that "neither better organization nor technological innovations appear to make [such complex systems] any less prone to system accidents." The inextricability of complexity and failure, therefore, makes it impossible to argue that any one designer is "responsible" for failure. Unfortunately for Duncan and Gillum, their engineering predecessors had vigorously argued for licensing statutes that made them responsible by law for the entire project, without addressing whether they could ever hope to be responsible in practice.[49]

Failure Reconsidered

Over the years, then, engineers have made sense of failure in at least four ways. First, failures were seen as evidence of a lack of expertise and used to distinguish between competent engineers and incompetent charlatans, as was the case with mid-nineteenth-century failures such as the Dixon bridge. The ASCE used such collapses to argue that failure by definition was not a part of engineering. Investigations in this era tended to equate success and failure with competence and incompetence.

Second, failures could be seen as resulting from improper use of a design and could be used to shore up the authority granted to engineers, as with the Ashtabula bridge disaster. In this conception, failure was still not a part of engineering per se, but rather, resulted from the ways in which engineered products were used or misused. ASCE studies in the late nineteenth century continued to equate success and failure with expertise and ignorance, though now the qualifications of the users became as important as that of the designers.

Third, failures ranging from the Quebec Bridge collapses to the Tacoma Narrows Bridge failure, and the deaths on the Golden Gate Bridge and Empire State Building during construction, could be seen as the price to be paid for pushing the boundaries of knowledge and could thus be used to emphasize the cutting-edge, brave nature of engineering. Here we see failure begin to be viewed as part and parcel of engineering practice—the inevitable price of progress, a risk associated with the state of the art. Still, day-to-day engineering projects were viewed as safe. In this conception, success and failure were equated with the routine and the cutting-edge, respectively.

Fourth, failures could be seen as inherent parts of a complex, interconnected system, in which no one person could be held "responsible" for the entirety. The Hyatt walkways, Three Mile Island, and the Challenger accident might, then, be just that—accidents. Failures can thus be viewed as part of engineering: never desired, but never completely preventable, even in "routine" engineering. Perrow suggests that the cause of certain failures "is to be found in the complexity of the system . . . It is the interaction of multiple failures that explains the accident." Success and failure here are to be equated not with anything else, but with each other—success and failure are, in short, inseparable.[50]

Each of these four models makes sense in the context in which it emerged. Engineers in each era were responding to the particular types of failures with which they were faced. But none of the four seems quite appropriate to help us come to terms with the collapse of the Twin Towers.

Clearly, in response to the WTC collapse, engineers and the public quickly rejected the first notion, that the designers had somehow been less than competent in their roles. Although a few hinted that Leslie Robertson or Minoru Yamasaki might be partially to blame (by not having made the towers even more robust than they were), by and large students of the collapse agreed that the towers were generally well built for their time.

From a certain perspective, the collapse of the Twin Towers could be seen as an example of second type of failure—improper use of a design—but not in the way that nineteenth-century engineers had envisioned it. The Twin Towers did not fail because they were poorly maintained or infrequently inspected. Although it is possible to say that the towers were "overloaded," it seems necessary to distinguish between the sort of overloading that results from sending heavy trucks over a bridge meant for light vehicular traffic, on the one hand, and overloading that results from someone intent on destroying that bridge. In the former case, the misuse still involves "use"; that is, the bridge is still being employed for a variant of its intended purpose. In the latter case, "misuse" seems a misnomer, since destruction—or prevention of use—is the intent.

The third interpretation of failure—that it is the price of progress—also seems hardly tenable here. It is difficult to imagine arguing that the nearly 3,000 deaths on September 11 were simply the price of cutting-edge skyscraper design. If one were not focused on engineering, it might be possible to suggest that risk of death is the price of freedom, but such a view does not directly help us to make sense of why the towers fell, nor how engineers ought to respond.

The fourth conception of failure is perhaps closest to the mark here but raises difficult ethical questions. September 11 did represent "system failure" of a sort: airline security, immigration policies, and national intelligence all contributed to the crashes, which in turn led to the collapse of the towers. Most of us would hesitate to accept that the terrorist attacks should fall under Perrow's category of "normal accidents," however. Perrow's analysis of what he termed the "social side of technological risk" did not include the sociology of terrorism, and it thus seems a misapplication of his theories to attempt to use them to make sense of the Twin Towers' collapse.

As we saw in Chapter 1, part of the lesson of September 11 lies in accepting that success and failure are not black-and-white polar opposites. To accept such ambiguity, such interconnectedness of success and failure, should not result in our simply accepting failure as a given, however. Finding hope in the ruins requires sitting with the discomfort of knowing failure is possible, even when we have done our best work; yet finding hope also requires that we continually strive to reduce failure, to be open to learning from it when it happens, and to make our next attempt better than our last. In Chapter 3, we turn to a group of engineers attempting to find hope in the ruins of the Twin Towers by exploring how to move their profession to a new era.

FOR FURTHER EXPLORATION

Select a recent disaster (e.g., the levee breaks in New Orleans, the Minneapolis I-35 bridge failure, the Paris airport collapse) and find two articles or essays about it, one from a popular media source and one from an engineering publication. How do the sources talk about the failure and who was responsible for it? Is there any difference between the two descriptions?

CHAPTER THREE

"A New Era"

The Limits of Engineering Expertise in a Post-9/11 World

Sept. 11 wasn't really about buildings.

RON KLEMENCIC

To point out . . . that the terrorists were responsible is both accurate and unhelpful.

NEW YORK TIMES EDITORIAL

On April 5, 2005, the National Institute of Standards and Technology announced the findings of its own, more thorough and long-delayed version of the FEMA/ASCE study. Gene Corley approvingly noted that the "results of the NIST team's extensive study are in close agreement with the findings of the FEMA/ASCE study." Engineers had long valued such investigative studies for their ability to illuminate lessons for future practice. The value of such studies, according to one ASCE official, was twofold: "We both validate longtime practices and, on occasion, change long-held beliefs." These lessons helped provide some positive results from an otherwise devastating event. The ASCE official went on to explain that "catastrophic events like the collapse of the twin towers are rare, and the knowledge that engineers gain from them is essential." A colleague had noted in the weeks after the collapse: "It's important that we look at the World Trade Center as an opportunity to make buildings safer and learn from them."[1]

As we saw in Chapter 2, this desire to find a silver lining in the midst of the darkest clouds of disaster has long been a part of the American engineering mindset. And yet even while these studies were still under way, the profession found itself wrestling with an uncomfortable truth—that, in the words of one prominent civil engineer, "our knee-jerk reaction as an industry is to worry about fixing the buildings, [but] Sept. 11 wasn't really about buildings, it was about terrorism and airplanes." If 9/11 hadn't been about buildings, what lessons were engineers to draw from that day? And yet, as the *New York Times* astutely noted, "to point out that the terrorists were responsible is both accurate and unhelpful," for what then could be done better in the future? For a profession long dedicated to learning from failure, 9/11 posed a crisis, much as the Quebec Bridge and Kansas City Hyatt Regency disasters had done, that required a new way of thinking—about failure, about responsibility, and about the limits of engineering expertise.[2]

As the investigations continued, building professionals and others began to discuss how best to respond to the attacks and more specifically to the collapse of the towers. Henry Petroski noted soon after 9/11 that the towers' collapse "signaled the beginning of a new era in the planning, design, construction, and use of skyscrapers." But the exact nature of this "new era" remained unclear. How, if at all, should building practices change in the future? Should existing buildings be retrofitted and, if so, which ones and in what ways?[3]

Design and Construction in the Aftermath of Disaster

Two events in the fall of 2001 provided forums for construction professionals to discuss their concerns and their visions for the future. The first gathering took place in Chicago in mid-October, hosted by the Council on Tall Buildings and Urban Habitat, an interdisciplinary group concerned with "all aspects of the planning, design, and construction of tall buildings." Just over two dozen architects, engineers, developers, security consultants, and others attended. The second meeting took place three weeks later in Washington, D.C., this time sponsored by *Engineering News-Record*, the leading news magazine of the construction industry. Some 69 invited industry leaders and over 200 attendees of the Design and Construction in the Aftermath of Disaster meeting came to discuss their concerns over the challenges facing their industry. The attendee lists for the two events overlapped, as many of the leaders in the field sought venues for exploring what the attacks would mean for the future of their profession.[4]

At both meetings, as each presenter offered suggestions for issues to consider and improvements to assess, it became clear that little consensus existed on the cause of the buildings' collapse, much less on how to respond. In part, this was because the investigations were still under way. But more important, the breadth of views reflected uncertainty as to how to categorize or think about this unexpected event. As transcripts of the discussions make clear, the participants were basing their suggestions for the future on five distinct explanations for what had occurred.

For some, the key cause of the collapse was the fire—an unprecedentedly huge one, but a fire nonetheless. The appropriate lessons, therefore, had to do with fire detection and suppression and building evacuation. It wasn't necessary to predict which buildings were at risk for future terror attacks, one attendee cautioned his colleagues: "We can't get too hung up on the terrorism aspect of the event. We had a building that had a major fire in it." This group argued that because fires could occur in any tall building, for any of a number of reasons, the lessons of 9/11 were applicable to any skyscraper: "We shouldn't address the issue of terrorism—preventing that in buildings. We can't. We can't do that. But what we can do is focus on making the buildings behave better in a calamity, whether an earthquake, fire, or plane crashing into it."[5]

The most dramatic change suggested by this group was a rethinking of the three-decades-old notion that "exiting [i.e., evacuating] an entire hi-rise building is not functional or feasible." The fire protection strategies that had been in place since the late 1960s had emphasized "orchestrated egress," evacuating tenants in order of their proximity to the fire, with the expectation that most of a skyscraper's occupants would never need to leave. Changing this strategy would have consequences for stairwell and elevator design, which would in turn affect the amount of rentable space in a building. In the meantime, it would be necessary to educate the public that "buildings are not designed for mass evacuation."[6]

For others, while the fire was important, it was outweighed by the role of the planes and more specifically airport security. Far better than dealing with a fire would be to "prevent these things from happening in the first place." It would be unrealistic to expect most buildings to withstand the impact faced by the World Trade Center and the Pentagon: "Stopping a plane in five feet is just not in the cards. . . . The good news is, if we understand this, buildings should not and cannot be designed for airplane attack. It's a problem, really, about airplane security." Robert Preito, chairman of Parsons Brinckerhoff, Inc.—one of the oldest and most prestigious engineering firms, with headquarters in New York City—concurred that it was necessary to "look at security from a systemic standpoint . . . the security failing was not at the World Trade Center or the Pentagon, it was at an airport."[7]

In other words, the lessons of 9/11 did not have to do with structural engineering at all. "There is no defense for somebody flying an airplane into a building, whether it's the World Trade Center or the Pentagon. There is no defense. That should not be the focus of our discussions nor should it be the focus of our society." By the time the planes were on their way into the buildings, it was already too late. As one attendee summed up bluntly, "It's not the building's fault, it's the plane's fault."[8]

For a third group, the planes were simply the weapons and the fires merely a symptom; the terrorists themselves were the root cause, and focusing on the fires or the planes or even the buildings missed the central point. The nature of the threats faced by buildings and their occupants was now fundamentally different: "Tall buildings have always been designed to withstand many potential hazards, fire, wind, seismic, weather, energy use, et cetera. I think we do have to consider malicious acts are also a hazard and must now be designed for." If terrorists were intent on striking again, they would find a way to do so—protect against airplanes, and terrorists would turn to anthrax and other biohazards.[9]

From this perspective, it was necessary for engineers to think explicitly about the nature of the threat posed by terrorists. A representative of the security industry argued: "I agree we shouldn't overfortify our buildings, but in some cases there are different considerations. We have to be conscious of the weaknesses and Achilles' heels of our buildings in the world of terrorism." But it was also necessary to recognize that even reinforcing both buildings and airport security would not be enough; building "concrete fortresses in the sky isn't our ultimate answer." Indeed, as one leading engineer in attendance noted, the buildings themselves were hardly the point: "We need to think about the hazards in terms of, I call it the demand side, the side where the evil originates. We need to focus on that, rather than on the capacity side, on trying to harden our facilities." Walker Lee Evey, program manager for the Department of Defense's Pentagon renovation cautioned his colleagues: "We spend a lot of time these days talking about what our response might be to a terrorist act, and what the targets might be, etc. . . . We're not spending very much time trying to understand the nature [and mindset] of our opponent. That's probably the first vulnerability we ought to address."[10]

A fourth explanation for the collapse argued that the symbolism of the buildings had drawn the terrorists to them as targets, and therefore was the key issue—"that's the source of the danger, the icon value or the occupant that's in the building." Any efforts to design against catastrophe in the future would need to recognize that it would be impossible to reinforce every building in the United States; instead, risk assessment would be the key: "You can't treat all buildings the same." The terrorists had not chosen the Twin Towers and the Pentagon at random. Which buildings or structures would be most at risk in the future? "If you look at how we've designed high-risk targets in the past, whether they were . . . defense facilities or nuclear plants, we used risk-based design factors. We didn't try to design for everything. We made assessments of what the threats were, what the probabilities were, and developed appropriate design criteria. September 11th has expanded the range of facilities that I think you want to consider in this way."[11]

And finally, a small group noted hesitantly that international relations and U.S. energy policy were the fundamental catalysts for the attacks and their outcome. As one attendee suggested, "The dependency on oil is part of the problem." But if the source of the problem lay in a nation's addiction to foreign-sourced energy, what help was that to the U.S. construction industry, looking for future directions?[12]

There's a powerlessness evident in the commentary at these events that is clearly uncomfortable for a profession whose members felt they ought to have the answers to solve the problems of the day. What the attendees at the Council on Tall Buildings were wrestling with was more than simply the semantics of how to discuss the largest disaster any of them had ever seen. They recognized that how they decided to categorize what had happened on September 11 would shape what was to come and indeed would determine whether they were even the appropriate people to provide the needed solutions.[13]

Note, too, that in the discussions these building professionals were doing more than trying to identify the root cause of the collapse; the heart of their debate is really about what cause will lend itself to the most practical and effective solutions. Clearly, on some level they all agreed that the buildings would not have fallen without the actions of some 19 individuals. The real question, then, was not so much a factual question of identifying the root cause, but a judgment question of selecting the most tractable cause.[14]

Cause and Effect

The issues that these building professionals were tackling were not entirely unique to the Twin Towers' collapse or even to engineering. Kenneth Carper, a leader in forensic engineering research, made a distinction between the technical explanation for a failure and the underlying reasons for that failure:

> When failures are discussed in professional journals, the typical article focuses on the technical/physical cause of the failure. There is a need for more discussion of procedural issues. There is always a technical/physical explanation for a failure, but the *reasons* failure occurs are often procedural. . . . Procedural causes are usually interdisciplinary, involving communication deficiencies and unclear definition of responsibilities."[15]

In the aftermath of the Chicago heat wave of 1995, for example, which killed some 739 people, one scholar noted that although the weather seemed the obvious culprit, it "accounts for only part of the human devastation that arose from the Chicago heat wave. The disaster also has a social etiology, which no meteorological study, medical autopsy, or epidemiological report can uncover." Those who died had clearly succumbed to the heat, but the *reasons* for their deaths were more complicated. Eric Klinenberg argued that to understand the full picture of the deaths in Chicago that summer required exploring "how the nature, culture, and politics of the city crystallized in Chicago in the summer of 1995." Low-income, elderly, solitary African American residents of Chicago were far more likely to die during that searing week, suggesting that race, class, and access to resources and services were as instrumental in their deaths as was the soaring temperature.[16]

If city officials assumed that Mother Nature or an act of God had brought the heat wave and with it the hundreds of deaths, there was little they could do beyond offering compassion to their fellow Chicagoans. What Klinenberg wanted them to understand was that however unavoidable the heat wave itself may have been, the consequences could have been significantly shaped by municipal policies and actions, such as ensuring that residents in poorer neighborhoods (where air conditioning was rare) and older residents (who were more susceptible to the heat) had ready access to shelters, water, and medical care as the temperature climbed.

The lesson of the Chicago heat wave, then, is that even when the cause of a disaster lies outside your control, the effects of that disaster may be at least partly in your hands. The building professionals who were discussing the collapse of the Twin Towers in the fall of 2001 were similarly wrestling with how to distinguish between cause and effect. Engineers had not been responsible for the attacks of 9/11 any more than Chicago officials had been responsible for the heat wave that gripped their city, but which effects *could* they have mitigated? And which ones, if any, *should* they have mitigated? In other words, apart from what was technically possible for engineers to do, what was morally appropriate to expect of them?

Engineering professor Charles Fleddermann provides a useful framework for discussions of disasters and engineers' responsibility for them. Fleddermann describes three types of "accidents"—engineered, procedural, and systemic—and proposes that engineers should think seriously about the extent of their responsibility for preventing each type.[17]

"Engineered accidents," such as the DC-10 airplane crashes of the 1970s and 1980s, result from design flaws. The design of the cargo doors on the DC-10 allowed the doors to appear to be securely latched when in fact they were not. At cruising altitude, the pressure inside the cargo hold pushed out on the cargo doors; if they were not secure, the doors could be blown off the aircraft and the resulting pressure differential between the cabin and the cargo hold would buckle the floor. In the DC-10 design, this buckling resulted in damage to the primary and backup electrical and hydraulic control lines running through the floor. The design of the locking mechanism on the DC-10 cargo doors was inherently more prone to trouble than the cabin doors of the DC-10 or the cargo doors of other aircraft and, according to one analysis, "produced an accident waiting to happen." In the case of one Turkish Airlines flight, 346 passengers died after an improperly latched door blew off in flight, severing all the control lines and leaving the pilots helpless. Because the weakness of the DC-10 arose from this flaw in the door's design and from the lack of redundancy in the controls, this type of accident is, according to Fleddermann, clearly in the realm of engineers' responsibilities.[18]

"Procedural accidents," such as plane crashes caused by pilot error, result from misuse of a product. The users of the product shoulder the majority of the responsibility for such accidents, but Fleddermann suggests that engineers consider ways in which their designs might make them prime candidates for misuse. A confusing or hard-to-read cockpit control layout, for instance, could result in a plane crash whose proximate cause is pilot error, but whose underlying cause is the engineers' failure to refine the controls, or a decision to refine them at an inopportune time. Designer Steven Casey describes a World War II pilot racing to take off from his airbase before incoming Japanese fighter planes arrived. As his colleagues found their planes and took off, this pilot jumped into the cockpit of the one new P-47 plane at the base and realized,

> Something was not right. Something *really* was not right. Jesus! The whole cockpit was different! It *couldn't* be all that dissimilar from the earlier models. All he needed to do was catch his breath and sort things out. . . . His shoulders were hunched forward, his eyes glued to the unfamiliar control panel. He couldn't think. It was just too much to fathom. . . . It was hopeless. There was no way in the world that he was going to get this thing up in the air and survive the aerial attack.[19]

The altimeter, the fuel gage, the ignition switch—all had been relocated. The pilot's training, experience, and instincts were of no use as he struggled to take off in this new and unfamiliar cockpit under stressful conditions. In the end, he managed to evade the Japanese pilot by racing his new plane around on the ground, "but he never did figure out why someone would redesign a fighter plane's instrument panel in the middle of a war." There was nothing inherently wrong with the new layout, except that the timing of the change presented an unnecessary challenge to pilots used to the earlier design.[20]

"Systemic accidents," according to Fleddermann's definition, result from a combination of errors occurring in various parts of a system, such as happened with the 1996 ValuJet crash in the Florida Everglades. Typically, any of the errors that occurred prior to Flight 592's takeoff would not have resulted in failure had they happened in isolation, but in combination they were fatal. The problems began with a box of oxygen canisters that had been removed from another ValuJet plane and loaded in Flight 592's cargo hold. Maintenance crews failed to install safety caps, which are intended to prevent accidental explosion of the canisters, and a shipping clerk then improperly packed and labeled the canisters, increasing the chance of accidental firing. The ramp agent and co-pilot both mistakenly allowed the mislabeled and hazardous cargo on the passenger flight. The cargo handlers placed the canisters, which require good ventilation to avoid accidental explosion, atop a set of plane tires in the hold, which added dangerous fuel to the fire that ensued when the canisters did explode. And finally, the cargo hold did not have heat or smoke detectors, which left the pilots with insufficient time to address the problem growing beneath their feet. As Fleddermann notes, "By themselves, none of these lapses should have led to the crash. However, the convergence of all these mistakes made the accident inevitable."[21]

Here, too, Fleddermann argues that engineers have a certain responsibility. Given the preponderance of engineering designs that are systemic, engineers ought to make a habit of considering the interactions of the system's components and how such interactions might result in dangerous consequences. Engineers cannot be expected to prevent all such failures, but that does not relieve them of the obligation to consider the possibility of such failures.

Fleddermann's nomenclature is helpful in categorizing failures, and his labeling of failures as "accidents" reinforces the reasonable assumption that failures do not occur because of any maliciousness on the part of engineers. Even if we allow that engineers in general do not wish to see their designs fail, the attacks of September 11 make clear that there are those in the world who may use an engineer's designs for evil rather than for good. It is one thing to suggest, as Fleddermann and others do, that engineers should protect the public (even the idiots among us) from hurting ourselves. It requires a significantly broader leap to suggest that engineers should protect the public from harm by terrorists.[22]

Given our naiveté prior to September 11, we can argue that the WTC engineers were surprisingly prescient in considering the effect an airliner would have were it to strike the towers. We can, in retrospect, be grateful that those engineers went above and beyond what was expected of them. But in a post-September 11 world, can engineers continue practice as usual? Or do the terrorist acts require a reconsideration of the meaning of "holding the public safety paramount"? Do we need to add a fourth category to Fleddermann's list of accident types? Intentional, deliberate acts that result in failure are not "accidents" in our usual understanding of the word, but they are, we must now admit, a potentially serious threat to engineering designs.[23] Should engineers now be expected to predict all of the malicious ways in which their designs might be used?

Civil engineer Aarne Vesilind thinks so: "A response to this [new, post-9/11] threat requires a two-pronged approach—technical and social." Vesilind argues that technological improvements will "no doubt have a beneficial effect," but he also believes that "better technology cannot be the only, or maybe not even a primary, engineering response to the threat of terrorism." According to Vesilind, if we are to combat terrorists' intentions, "we have to develop a greater knowledge of diverse cultures and societies and ask why people would want to cause harm, what their motives are, and what drives their actions." This knowledge cannot come from the equations and laws and problem sets of a technical engineering course.[24]

Just as we can categorize the causes of failure, so too can we categorize the lessons we take away. First and most narrowly, one can learn technical lessons applicable to the current design or project (as might be drawn, for example, from the FEMA report). In the case of the Twin Towers, such lessons were largely moot, as there was little likelihood that the towers would be rebuilt. The rebuilding of the Pentagon, however, did benefit from lessons learned about the effects of the airplane on that structure. Most often, these narrow lessons come during the design process itself, when engineers test their prototypes, often to the point of failure. The information gained from such failures is important but by definition not broadly applicable.[25]

Second, one can extrapolate how those lessons might be applicable to other, similar designs in the future. In Manhattan, the rebuilding of WTC 7 across the street from the Twin Towers reflected lessons learned about WTC 1 and 2, including fireproofing, stairwell placement, and evacuation procedures. The most famous historical example is perhaps the Tacoma Narrows Bridge—the famed Galloping Gertie caught on film in 1940 as it tore itself to pieces in the wind. Engineers learned invaluable lessons about harmonic oscillations and the aerodynamics of thin bridges. The ASCE has published a volume entitled *Failures in Civil Engineering* summarizing roughly 50 failure case studies and providing a brief overview of the lessons learned following each disaster. The lessons cited range from narrowly technical ones (such as a miscalculation of likely wind forces, as in the Tay Bridge collapse of 1879) to broader developments (such as the recognition of the need for additional study of aerodynamic effects on suspension bridges, as in the Tacoma Narrows case).[26]

Third—and here lies the biggest challenge—is to imagine lessons that go beyond obviously similar projects and are applicable to design in general. It is unlikely that the events of 9/11 will be replicated. As Robert Prieto cautioned his colleagues at the Aftermath of Disaster conference, "We have to be careful that we don't get trapped into designing for the last threat when the next threat is out there."[27] The challenge, then, was how to extract lessons from a specific event of the past that would prove useful in the as-yet-unknown future. This is another of those contradictions that define the profession; engineering requires depth of knowledge, but breadth of application.

We spent a great deal of time and money as a nation ensuring that boxcutters could not be brought on board airplanes again, but the next terrorist attack will doubtless use a different method. It is, as the saying goes, like closing the barn door after the horse has escaped. What broad lessons can be drawn from the collapse of the Twin Towers, then, that will improve safety? Ideally, the lessons will be sufficiently broad as to be beneficial even if no additional terrorist attack is imminent. It is useful to remember that the WTC engineers were unable to foresee the attacks of 9/11 for the same reason that we are unable to predict with precision what the next threat will be. The best lessons we can learn, then, are those that will help protect us in situations whose cause we can't imagine but whose occurrence would be devastating.

One way to brainstorm is to find an analogous case for exploration. For our present purposes, a useful analogy lies on the floor of the North Atlantic. It might seem a strange place to look for a parallel to the design of a modern skyscraper, but even a brief analysis reveals instructive comparisons between a record-breaking and supposedly unsinkable ocean liner from the early twentieth century and a record-breaking and supposedly indestructible skyscraper built a half-century later.

Fire and Ice

In 2005, two *New York Times* journalists completed a careful, minute-by-minute account of "the fight to survive inside the Twin Towers" and became convinced that the performance of the buildings was more complicated than the ASCE/FEMA study team's findings suggested. Jim Dwyer and Kevin Flynn took the *Titanic* as a model for understanding what had happened on 9/11, arguing that similar flaws and overconfidence plagued the Twin Towers. In both cases, little attention had been paid to evacuation or rescue because so much faith had been placed in the infallibility of the structure. In theory, both had been designed so well that wholesale evacuation would not be necessary.

The *Titanic* had had more lifeboats than required by law, but fewer than required by its capacity. Similarly, the Twin Towers had as many stairwells as called for in the New York City building code of 1968, but fewer than necessary for evacuating tens of thousands of tenants. The Twin Towers' designers and planners had also chosen not to include "fire stairs"—specially reinforced and insulated stairwells with double sets of doors at landings to limit the spread of smoke and fire. The existing stairways were clustered in the buildings' cores, to maximize rentable space. While this made for an efficient design, it also made for efficient destruction, as the hijacked planes destroyed five of the six staircases in a matter of seconds, trapping the vast majority of those at or above the crash sites.[28]

Although the *Titanic* succumbed to flooding and the Twin Towers to fire, the parallels continue. The *Titanic*'s hull had been designed in sections, so that if one flooded, the water would be contained and would merely damage, not devastate, the ship. A similar approach in the Twin Towers intended that any fire that broke out would be restricted to one or two floors and could either be extinguished or allowed to burn itself out without spreading to other areas of the building. A major fire in the North Tower in 1975 had indeed been confined to just a few floors, and though there was some structural damage, it was limited to a small area, and did not threaten the integrity of the structure overall.[29]

As with the *Titanic*'s flooded compartments, however, the Twin Towers' fire containment system failed quickly on 9/11 and with devastating consequences. Because the designers had not anticipated a fire this size, the firefighters had limited options and knew from the first minutes that theirs "would be strictly a rescue operation" rather than an attempt to douse the flames. As Dwyer and Flynn explained, "The FDNY could fight a fire on one floor, maybe two. They could not handle what confronted them now—at least five floors fully engulfed." Even if water had been available near the site of the blaze, firefighters could not hope to tackle an inferno of this size, but instead would have to hope it burned itself out while they "concentrate[d] on helping people evacuate." And given how high the fires were and the fact that the elevators were almost entirely out of commission, the time it would take crews to reach the fires would be "measured in hours, not minutes."[30]

Thus, physical aspects of the buildings limited the options available to evacuees and firefighters. A less tangible but equally crucial factor would further hamper their efforts: communication. That communication problems hindered the evacuation was not directly an issue of "building performance" but does provide evidence of the complexity of understanding the events of the day. Tenants, especially those trapped in elevators or stairwells, had little clue what had happened to the buildings they were in. The few who were able to reach friends and family by telephone learned more of their predicament than was known by the police and firefighters sent to rescue them, who could see little from their command post in the lobby of the North Tower and who had difficulty transmitting by radio within the towers. To make matters worse, police and firefighters were largely unable to communicate with each other, so key intelligence gained by one group could not be readily shared with the other.

The communication gaps resulted in at least some of the deaths that day—of tenants high in the North Tower unaware that Stairway A remained a viable escape route, and of firefighters whose radios could not pick up the evacuation orders issued by their commanders nor those broadcast by the police. News of the South Tower's collapse had barely penetrated the North Tower before it, too, fell.[31]

Dwyer and Flynn pointed out that the terrorists themselves had never anticipated the buildings' collapsing and argued that although the terrorists were culpable, "the buildings themselves became weapons" that day. Of the 1,500 or so souls who survived the initial crashes but died in the building, Dwyer and Flynn declared that "those people were not killed by the planes alone any more than passengers on the *Titanic* were killed by the iceberg." They went on to explain the myriad ways in which the buildings' design and "a sclerotic emergency response culture in New York" aided and abetted the terrorists:

> Those who did not escape were trapped by circumstances that had been the subject of debates that began before the first shovelful of earth was turned for the trade center, and that continued, at a low volume, through the entire existence of the towers. Could the buildings withstand the direct impact of an airplane? Was the fireproofing adequate? Were there enough exits? . . . The fate of all the men and women inside the towers during those 102 minutes [that the towers remained standing] was specifically, and intimately, linked to decisions about the planning and construction of, and faith in, colossally tall buildings.

The 1,000 or more people trapped in the upper floors of the North Tower had survived the brunt of the terrorists' attack but would die because they had no way to escape, thanks to decisions that had been made before most of the hijackers had been born: "Their fate was sealed nearly four decades earlier, when the stairways were clustered in the core of the building, and fire stairs were eliminated as a wasteful use of valuable space." In short, Dwyer and Flynn declared, "the structural engineers of the trade center had anticipated that the towers would be able to respond to the stress of an impact from the airplane. No one had designs, however, for the people inside."[32]

If one builds a tall building or a large ship, it takes little imagination to suppose that it might be necessary to evacuate that structure. The lesson of the *Titanic* was not that one should avoid icebergs (which was clearly known before that, although additional precautions were put in place after April 1912), nor even that ship design ought to be more robust. Rather, in hindsight, the lesson of the *Titanic* was that no matter how "unsinkable" you believe your ship to be, you should still equip it with enough lifeboats for all the passengers and crew on board, and you should still have a plan for an organized, efficient evacuation. Or, more broadly still, in designing any structure that will be occupied by large numbers of people (ship, skyscraper, or stadium), one should think about how all those people can be quickly removed from that structure in case of an emergency. Are there enough exits? Do people know where they are and how to use them? Is redundancy built into the structure?[33]

This is not to say that the World Trade Center failed in evacuation: clearly, a great many people escaped (though if the towers had been fully occupied that morning, evacuation would have become far more difficult and time-consuming). Rather, where did difficulties arise? Improved communication during the evacuation would have let occupants know whether to stay or go, or what paths were open to them. Instead, one announcement in Tower 2 encouraged occupants to return to their offices, and in the same tower, some tenants headed to the roof, where evacuation was not possible, while the fact that Stairway A remained open was largely unknown to those trying to escape.

One lesson of the *Titanic*, learned by Senator William Alden Smith, who chaired the U.S. investigation of the great ship's sinking, was that in spite of the shocking behavior displayed by individuals ranging from shipbuilders to wireless operators to ships' captains, no one had broken any laws, for the *Titanic* was built and sailed under laws written before her time. One writer about the *Titanic* noted, "The *Titanic* duly mirrored her culture. Growth in the size of things had exceeded the rate of accompanying changes in law. . . . In short, what *really* caused the loss of the *Titanic* were the unrecognized weaknesses of her day—hype, haste, and hauteur."[34] In some ways, the same could be said of the World Trade Center.

We might conclude, as an observer of the *Titanic* disaster did, that laws "have not kept up with progress" and that outdated legislation regarding lifeboats (or stairwells, in the case of the World Trade Center) is to blame for the calamities that result.[35] But to act as a professional requires more than simply following the law; a professional, by definition, ought to know better. Ethics codes call on professionals to practice conscientiously, to the best of their ability, and in the public interest, not merely to abide by the law, which is taken as a given and a minimum standard.

Sharing Responsibility

As we push the boundaries of the lessons to be learned from failure, new ethical questions are bound to emerge. For instance, if you cannot ensure safe evacuation, should you build the structure in the first place? Firefighters opposed the Twin Towers long before they were completed because they have no efficient way of fighting fires higher than their equipment can reach, over broad open spaces such as the one-acre floors of the World Trade Center. Defenders of the towers and their designers noted that no one could have anticipated hijackers flying fuel-laden planes into the towers at high speeds. But was such an imagination necessary? First, planes had hit skyscrapers before, and some Manhattanites had expressed concern that the same would happen to the Twin Towers. Second, it is not impossible to imagine that in a pair of skyscrapers the size of the Twin Towers something might go wrong—fire, accidental explosion, lightning, hurricane, earthquake. Indeed, some of these events were considered during the design. Whose obligation is it to raise and respond to such questions?[36]

One might reasonably object that engineers should not (indeed cannot) be responsible for all of this. In part this is true: there are lessons to be learned that will go beyond what engineers can or should handle—it may be, for instance, that the lessons to be learned from the WTC collapse have more to do with airport security than with building structure. But two cautions are necessary before engineers can abdicate responsibility: first, it takes a broad view to identify the best lessons to be learned (so engineers should be prepared to take such a broad view), and second, engineers have been responsible for an increasingly wide array of concerns and therefore should not assume that the responsibilities in the future will be the same as those of the past.

As necessary and productive a step as the FEMA/ASCE and NIST studies were, they would be of limited use if they remained in isolation. Engineering author Louis Bucciarelli explained the dangers of lessons narrowly learned: "Setting technology apart seems necessary if we are to make sense of it. We learn by categorizing as different all the bits and pieces of our experience. But in this business of classifying, it may be that the glasses we wear, the filters we set before us, or the sieves we use in collecting, sorting, dividing, and piling up our experiences and observations are not the right tools."[37] We have already noted the *New York Times*' comment that placing blame on the terrorists "is both accurate and unhelpful." To rely solely on a numerical, technical explanation of why the towers collapsed is also accurate and, by itself, unhelpful. As Ron Klemencic put it, September 11 wasn't only about the buildings.

And yet ignoring the buildings themselves and simply pointing the blame at the terrorists is equally limiting if we as engineers are looking for lessons to draw from this disaster. The point is not to assign blame but to be willing to think broadly and comprehensively about the causes and cures of disasters. We will understand what happened on September 11 only to the extent that we are able to reconcile the many answers to the question "Why?" and recognize that just as engineering cannot by itself explain the events of 9/11, neither can it in isolation offer solutions.

An important component of the engineering worldview is the long-held belief (or at least the claim) that—in the words of ASCE columnist Richard Weingardt—"practicing structural engineers have the answers for designing new structures and upgrading existing ones to withstand the forces tomorrow's world will require." Weingardt's column was addressed to a shaken profession in need of inspiration. With the benefit of hindsight and a more modest perspective, we would do well to amend his bold claim to say that engineers have *some* of the answers. This broad perspective and interdisciplinary approach lie at the heart of the "new era" that Petroski predicted would emerge after 9/11.[38]

FOR FURTHER EXPLORATION

1. In his testimony before Congress, WTC forensic investigator Gene Corley noted, "Because there is no limit to the destructive forces which terrorists can bring to bear against our built infrastructure it is impossible to design a building to withstand such an attack. The multi-faceted approach presently being pursued, that being to prevent the attack initially and pursue rational, scientifically based methods to improve structural performance, is both sound and prudent."[39] Find an article in the engineering or popular press about changes made since 9/11. Discuss which of the five explanations for the tragedy of 9/11 offered by the building industry at the start of this chapter are included in your article:
   - Building design
   - Airport/airplane design or security
   - Terrorist activity
   - Risk assessment (and differential responses based on that assessment)
   - International relations
   - Other?
2. Find a case study, article, or book about another disaster.[40] Compare and contrast the explanations given for the disaster and the responses suggested in order to prevent similar disasters in the future. Can you think of other possible explanations that would lead to different recommendations?
3. If you are engaged in a design project, identify all the people who are or should be included in the process (customers, fabricators, users, etc.). What role does each person or group play in each of the five stages of design (from Chapter 1)? Are there particular times when outside input is most important? Are there ways to substitute for such input? For example, can members of the design team effectively play the role of "user"?

CHAPTER FOUR

"Safe from Every Possible Event"

How to Strive for the Impossible

Surely, we have all learned the most important lesson—that the sanctity of human life rises far above all other values.

LESLIE E. ROBERTSON

Architect Minoru Yamasaki knew that the more open and flexible he could make his floor plans for the Twin Towers, the more desirable the buildings would be to prospective tenants. But how can one build 110-story towers without support columns scattered throughout each floor? The engineers' answer lay in concentrating the strength (and the weight) of the structure in the outer "tube," or perimeter, of each tower and an inner "core" that would house the elevators and mechanical equipment. The designers then added stability to these two components by tying them together via the lightweight trusses that would form the foundation for each floor of the towers and capping each tower with a massive hat truss spanning the roof. But design is so hard in large part because it is so interconnected. The closely spaced columns of the tube almost miraculously withstood the impact of the airplanes on September 11, but the tube-and-core design also made possible the lightweight floors and wide-open tenant spaces that could not resist the devastating fire that followed.[1]

In the preceding chapters we have taken as a given that safety is a goal. But good engineers must always test their assumptions to ensure their foundation will support that which is built atop it. One of the ideals of American engineering, expressed most clearly in its many codes of ethics, is its emphasis on the public's well-being. The first canon of the ASCE code of ethics, for instance, notes that "engineers shall hold paramount the safety, health and welfare of the public." In an essay written in the months after the collapse, WTC engineer Leslie Robertson wrote that no matter what else might come of the tragedy, "surely, we have all learned the most important lesson—that the sanctity of human life rises far above all other values."[2] It is true that the devastation of September 11 encouraged us all to remember that whatever else we might have lost that day, if we survived it with our lives, we were lucky indeed.

And yet to overemphasize the ideal of public safety and the sanctity of human life is to obscure a fundamental characteristic of engineering practice—that public safety is, in fact, never the paramount consideration. An inherent conflict exists between the nature of engineering (design under constraint, which requires tradeoffs) and the stated primacy of public safety. There are four reasons why public safety is not and can never be paramount in real-world engineering design.

1. Safety is never the only factor to be considered in engineering design and thus can never take priority over everything else. The very nature of design is that it involves a balance of many, often competing factors. The Twin Towers needed to be safe, but they also needed to be affordable and rentable or they would never be built.
2. Public safety is not paramount because the safety of each component may affect the safety of others or of the system as a whole. In other words, maximizing safety in one area routinely reduces safety elsewhere. Strengthen the tube and core of the Twin Towers, and of necessity the floor trusses must be lighter (and weaker) if the building is to stand under its own weight.
3. There is no perfect safety, just as there is no perfect strength or price or efficiency, or any other typical constraint on engi-

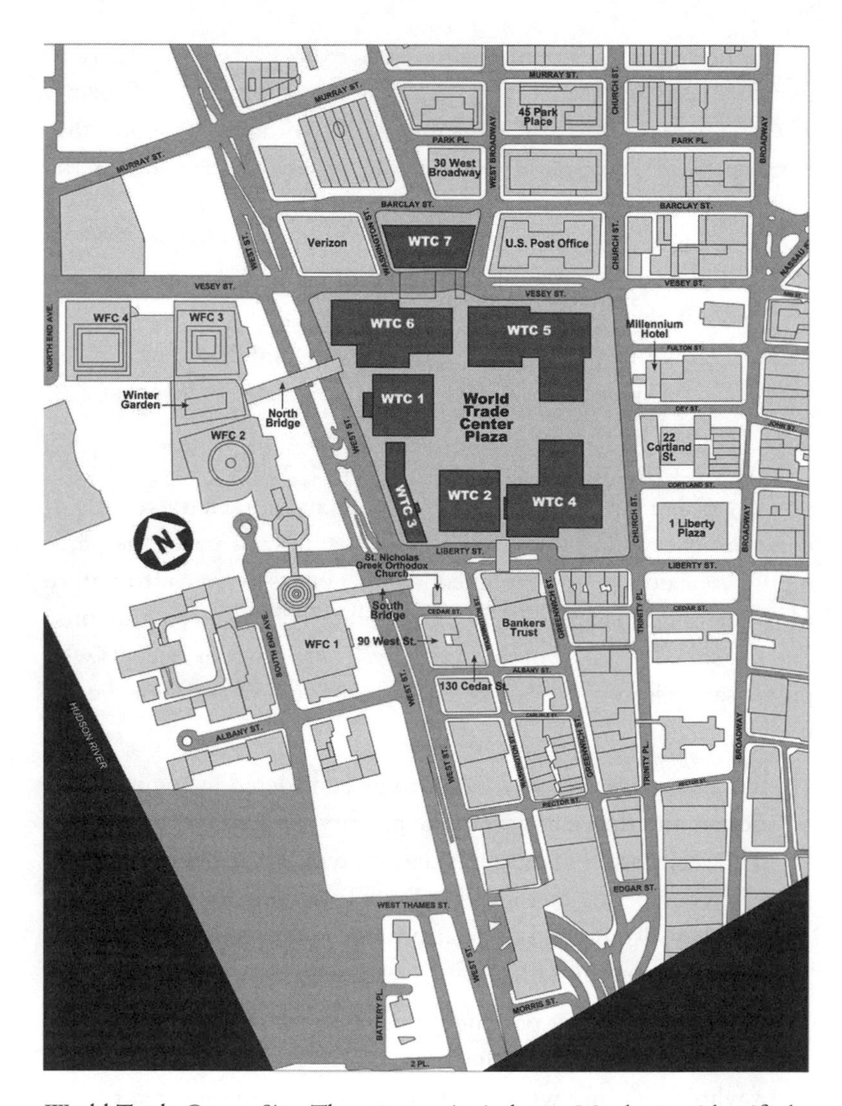

World Trade Center Site. The 16-acre site in lower Manhattan identified by the Port Authority as the location for the World Trade Center, into which the design team had to fit the requested structures. The World Trade Center, most notable for the Twin Towers, actually consisted of a total of seven buildings, each known by its number. The North and South Towers were known as WTC 1 and WTC 2, respectively. NIST, *Final Report on the Collapse of the World Trade Center Towers*, http://wtc.nist.gov/NCSTAR1/PDF/NCSTAR 1.pdf, p. 3.

neering design. Yamasaki's tube-and-core design was brilliant because it made the towers *acceptably* safe but also *relatively* inexpensive to build and *increased* the space available for rent. Just as the cost of a design can never be $0 and the space available for rent can never be infinite, risk can never be eliminated altogether and safety can never be fully maximized. The challenge for the designer is to determine what is enough in each of these areas.

4 Finally, as became painfully obvious on 9/11, the safety of a structure depends not only on the design itself, but also on how it is used (or misused). As important a role as the engineers played, the fate of the Twin Towers did not lie solely in their hands, or even in the hands of the entire design and build teams.

To say that public safety is never paramount is not to say that it is unimportant, or even that it is never a deciding factor in a design. Rather, to concede that public safety cannot be maximized to the exclusion of all other factors is to acknowledge yet another paradox in engineering: safety must be sought but can never be reached in the ideal.

Safety and . . .

By the time Minoru Yamasaki had been hired as the architect to design the World Trade Center, the Port Authority had already determined the location—16 acres in lower Manhattan, covering 13 city blocks, roughly half of which consisted of infill that used to be the Hudson River. By the time Yamasaki had accepted the commission, executive director Austin Tobin of the Port Authority had already decided that the World Trade Center would encompass 10 million square feet of office space. As Yamasaki's team contemplated possible models for housing such a vast amount of space, Tobin's engineer Mal Levy turned up to inform them: "The Port Authority wants this to be the most dramatic project in the world. It must be a symbol of New York. And we want the tallest building in the world." Ironically, it was also Levy who had earlier explained to his Port Authority colleagues: "We can't afford a corporate status symbol. . . . This is not a trophy we are building. It is a speculative office building." He set the cost of the building at no more than $22 a square foot, at a time when 40-story buildings in Manhattan were costing $35 per square foot.[3] Yamasaki's team had the outlines of their project determined for them before they ever began the real work of design. And Leslie Robertson, who would become the project engineer for the Twin Towers, had not yet entered the picture.

Safety was important, but there were additional practical matters at hand. Not only did Yamasaki and his team have to search for a way to incorporate all the many demands of their clients, but they also needed to recognize that solving one problem might well raise a problem elsewhere. Consider two such complications. The designers needed to design a structure light enough to rise 110 stories yet sturdy enough to stand under its own weight; and they needed to find a way to transport tenants efficiently through the building without taking up the entire structure with elevator shafts. Without solving these problems, the building would not be possible. Safety would have to come later.

Like many skyscrapers of the mid-twentieth century, the Twin Towers relied on steel framing coated in spray-on fireproofing rather than on reinforced concrete framing or steel framing covered in concrete. The use of metal for most of the frame reduced the overall weight of the structure (an essential consideration in a building rising 110 stories from the ground) and made the construction process easier and cheaper by reducing the amount of messy and time-sensitive concrete that needed to be trucked through the streets of Manhattan.[4] The choice made sense in terms of time and money, and also helped to increase the usable space in the towers. But those gains were offset by a reduction in safety, as metal is more vulnerable to fire than concrete is. Then again, a concrete-framed structure could never have risen to the height required of the Twin

Towers; if 110 stories were necessary, so too was an alternative to concrete.[5]

As with the choice of building materials, the design and placement of elevators in a high-rise building is no insignificant detail. Although higher floors may provide more expansive views and a more prestigious address, their cachet is lessened with every extra minute it takes workers and visitors to arrive. Yamasaki's team turned the concept of the city's subways on end and translated it to the skyscraper. Using a combination of local and express elevators, commuters could reach their offices on the highest floors without needing to stop at some 70- or 80-odd floors along the way: they could be whisked one- or two-thirds of the way to the top on an express, then change elevators at a "skylobby" and take a local the remaining few floors to their destination.[6] This streamlined design helped speed people up through the building while also minimizing the space needed for the elevators. But by concentrating all the elevators in the building's compact core, this system meant that damage to one elevator could well mean damage to many or all. Indeed, on September 11, the airplanes not only damaged many of the elevator banks, but most of the stairwells, too. Efficiency of use translated tragically to efficiency of destruction.[7]

The structural framing and the elevator system in the Twin Towers demonstrate that safety cannot be paramount because nothing is designed merely to be safe. Everything is designed to be safe *and* more—user-friendly, efficient, cost-effective, attractive, and so on. The Twin Towers designed with safety as the paramount concern would not have been the Twin Towers as we knew them. Indeed, it would be difficult to imagine them existing at all.

One mechanism engineers use to compare the relative strengths of design alternatives is a rating system called a design matrix (also known as a decision matrix or evaluation rubric), as shown in the table. The left-hand column of such a table lists design specifications or criteria: necessary square footage, height, and so forth (in the case of the World Trade Center, 10 million square feet, 110 stories, etc.). These specifications are often identified by the customer,

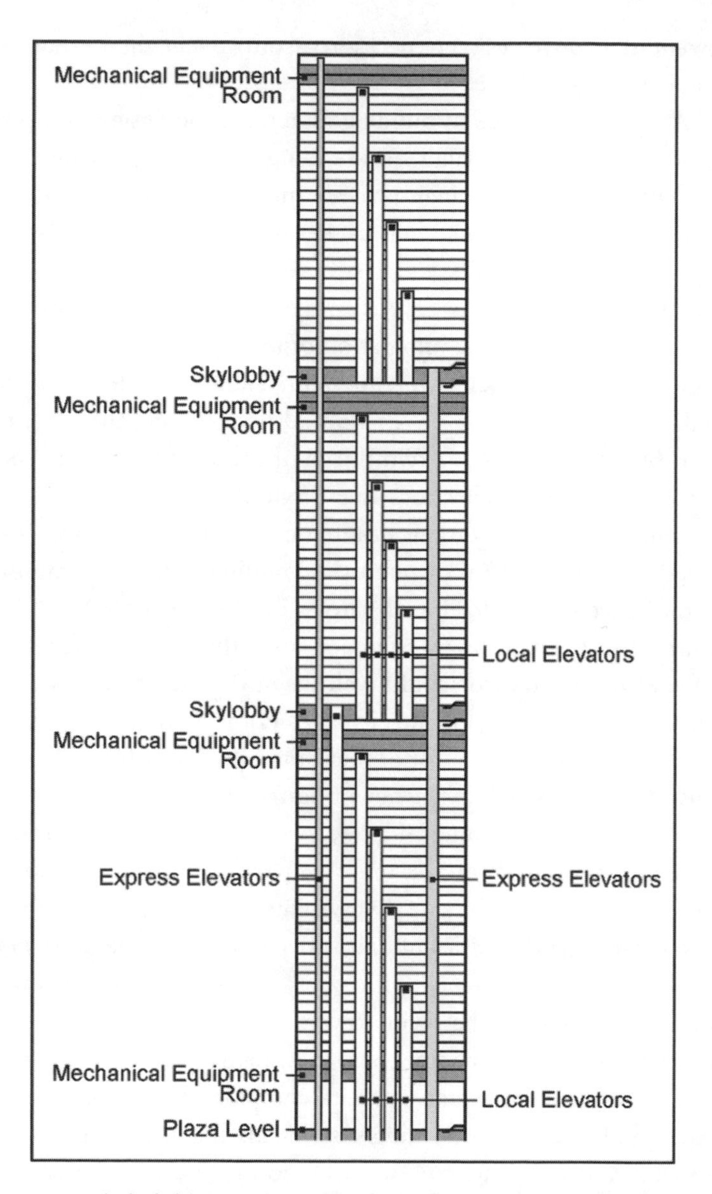

Elevators and Skylobbies. Schematic of the elevator layout in WTC 1 and WTC 2. Minoru Yamasaki modeled the combination of express and local elevators after the New York City subway system, with "skylobbies" in place of transfer stations. This system allowed him to minimize the square footage of each floor dedicated to the elevator shafts and maximize the amount of rentable open space. NIST, *Final Report on the Collapse of the World Trade Center Towers*, http://wtc.nist.gov/NCSTAR1/PDF/NCSTAR 1.pdf, p. 14.

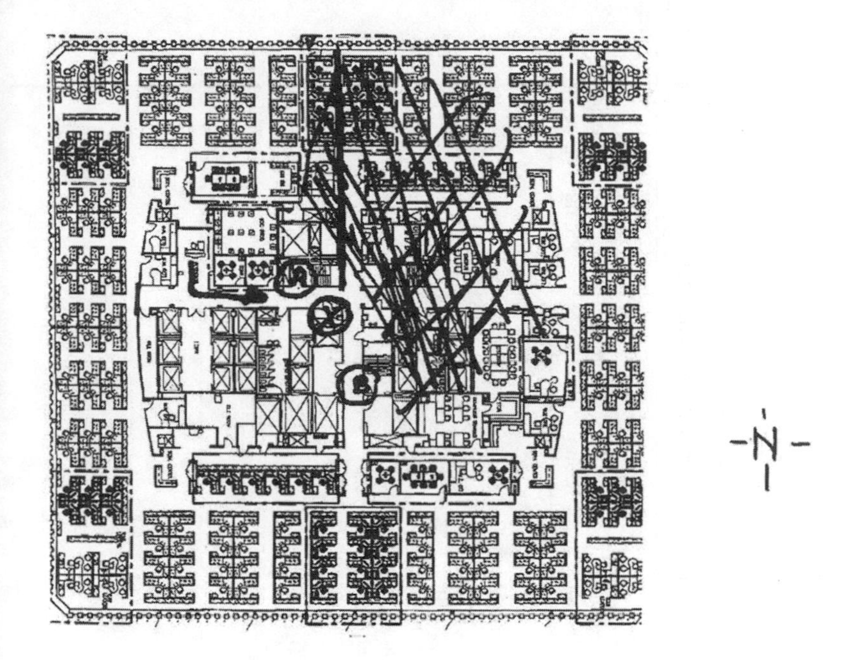

Floor Plan with Damaged Area Marked. This floor plan from WTC 1 shows the approximate area damaged by the plane as it entered the building. The immediate damage to evacuation routes was extensive and left no stairwells intact. The "open," or rentable, space on this floor was filled with workstations and offices. ASCE/FEMA, *World Trade Center Building Performance Study*, www.fema.gov/rebuild/mat/wtcstudy.shtm, p. 2-19.

but are quantified and weighted by the design team. The top row of the table lists several design options and may include a "control" or existing design against which to compare the options. The design team scores each design on each specification, and tallies a total score for each option. The resulting scores can then be used to support a decision to pursue one design over another, or to combine two possible designs.

Although the numbers lend an aura of certainty to the comparisons, the team must remember that the inputs are largely subjective and thus the outputs must be used with care. Note, for example,

Sample design matrix (also known as a decision or evaluation matrix) used to compare several design alternatives

		Design A (reference)		Design B		Design C		Design D	
Selection criteria	Weight (%)	Rating	Weighted score	Rating	Weighted score	Rating	Weighted score	Rating	Weighted score
Safety	Must[a]	✓		✓		✓		✓	
Ease of use	20	3	0.60	3	0.60	3	0.60	4	0.80
Accessibility	25	3	0.75	2	0.50	4	1.00	3	0.75
Usable space	10	3	0.30	4	0.40	4.5	0.45	3	0.30
Cost	10	3	0.30	3.5	0.35	2	0.20	2	0.20
Ease of assembly	15	3	0.45	2	0.30	2	0.30	2	0.30
Durability	15	3	0.45	1	0.15	2	0.30	3	0.45
Attractiveness	5	3	0.15	4.5	0.23	4	0.20	4	0.20
Total	100		3.00		2.53		3.05		3.00
Rank			2		4		1		2
Continue? (yes/no/combine)			Yes/combine		No		Yes/combine		Yes/combine

[a]Required

that in the sample design matrix shown here, "safety" has been listed as a "must" rather than being given a specific weight; this approach is appealing but obscures the fact that one design might be safer than another. In the case of a skyscraper, "safety" would likely be defined in more specific and quantifiable terms such as fire ratings, stairwell widths, and structural redundancies.

A Little More Here, A Little Less There

Suppose for a moment that one wanted to assess safety in the Twin Towers. One would soon have to confront the question of what it means to be "safe." We've noted already that the tightly spaced perimeter columns that so remarkably withstood being punctured by two 747s were accompanied by lightweight trusses and vast unobstructed spaces on the one-acre floors of each tower. Because the towers did not collapse immediately, thousands were able to escape from the buildings and the surrounding area, but the buildings were soon ravaged by devastating fires that assured their collapse and doomed occupants of the floors above the airplanes' impact. In gauging the safety of the towers, should one focus on the thousands who escaped or on the many who could not? Is it reasonable to suppose that more lives could have been saved from any structure so vast and so densely populated? Between them, the towers housed some 40,000 people—enough to populate a small city. Under what circumstances would we expect to be able to evacuate an entire municipality following a sudden accident or attack?

Some skyscrapers make use of so-called "refuge floors" spaced throughout the building that allow for a sort of on-site evacuation; occupants need only go five stories or so to the nearest such floor, where they are afforded enhanced protection from smoke and fire and can await safe exit from the building. Refuge floors would have provided increased safety for WTC tenants under many circumstances but could well have vastly increased the death toll on 9/11. As Leslie Robertson noted, "Thank God we didn't have ref-

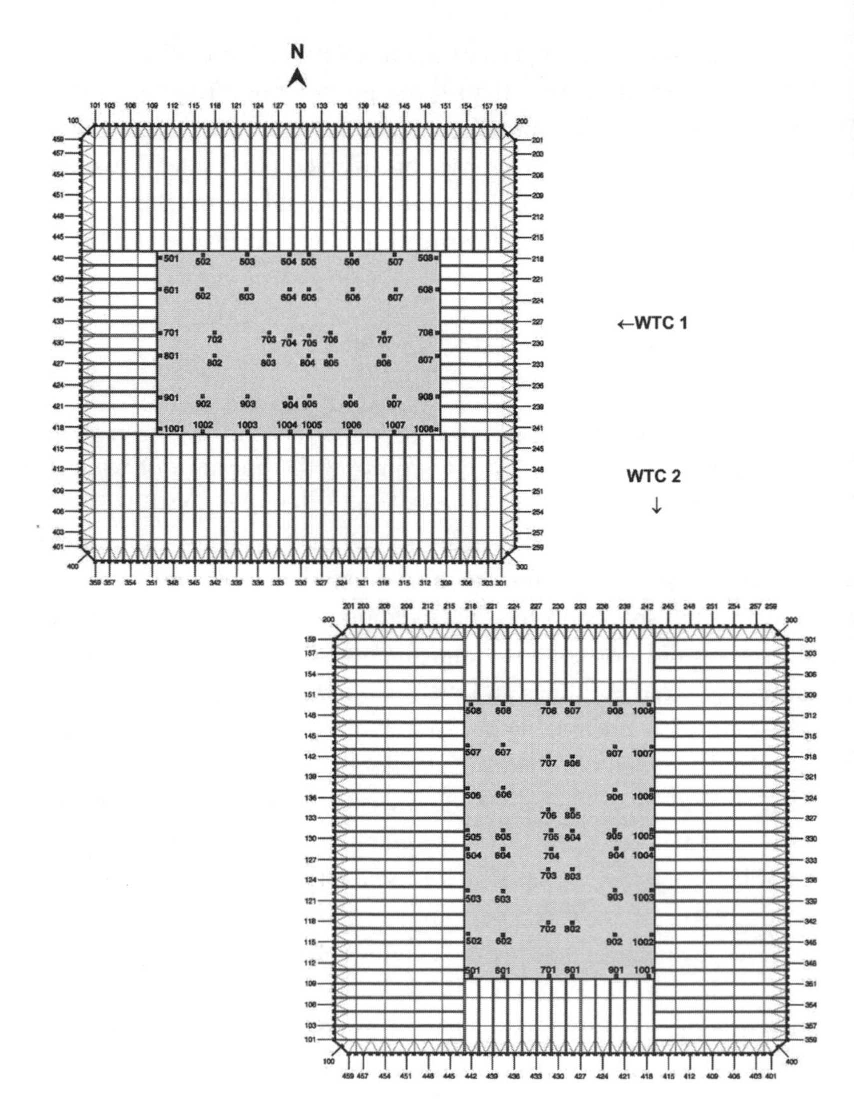

WTC Floor Plans. Typical floor plans for the North and South Towers of the World Trade Center. Although each tower is a square, the central cores are elongated. In Tower 1 (the North Tower), the core ran east-west, and the plane struck from the north, leading to less immediate core damage than in Tower 2, though it did sever all evacuation routes for those above the impact. In Tower 2, the core ran along the north-south axis; when the plane entered from the south, it did more severe structural damage to the core but left one stairwell intact. NIST, *Final Report on the Collapse of the World Trade Center Towers*, http://wtc.nist.gov/NCSTAR1/PDF/NCSTAR 1.pdf, p. 7.

uge floors in the World Trade towers—people might have stopped there to regroup, rather than getting out of the buildings as fast as they could."[8]

All of this is not to suggest that designers should avoid attempts to improve safety: however fortunate the lack of refuge floors might have been on 9/11, in other circumstances they have been lifesavers. Rather, the lesson here is that the complexity and interconnectedness of design require careful thought and an appreciation for the elusive nature of "safety." Safe for whom? For how long? Under what conditions and circumstances? At a professional conference for structural engineers some three weeks after 9/11, a fellow engineer asked Leslie Robertson, "Is there anything you wish you had done differently in the design of the building?" whereupon "Robertson broke down and wept at the lectern."[9] In the face of such an impossible question, it is difficult to imagine any other response, and yet engineers are faced with such decisions every day.

What Will Perfection Mean Tomorrow?

At some point in any project, the design phase must to come to an end and construction must begin. The specific timing of that turning point can turn out to be crucial in determining the safety of a design. For the Twin Towers, the construction phase came sandwiched between two important alterations to the definition of "safe."

The New York City building code had been updated in 1968, and although design of the towers was already well under way in that year, the Port Authority generously agreed to abide by the new regulations rather than continuing under the rules in place when Yamasaki and Robertson began their work. What a casual observer might not have thought to consider was that the new code was in fact more lax than its predecessor, which had been based on lessons learned in the deadly Triangle Shirtwaist factory fire of 1911.

A trade organization had long lobbied for a new code, on the grounds that the 1938 code was out of date and did not recognize or accommodate advances in building technology and encouraged over-fortifying buildings. The lobbyists argued as well that a code designed to apply to factories was inappropriate to the needs and (lower) risks associated with more mundane office space. The 1968 code for office buildings thus reduced the number of hours that columns and floors needed to withstand fire, and eliminated the requirement that structural steel be wrapped in concrete or masonry, in favor of allowing architects to determine the most appropriate method of fireproofing. The *New York Times* reported that under the new code, a wall "can be made of brick, specially treated wood—or shredded wheat—so long as it can resist fire."[10] Perhaps most significant of all for the Twin Towers, the new code eliminated the requirement for "fire towers" (specially vented and reinforced stairwells to facilitate evacuation during a fire), reduced the required number of stairways, and minimized the protections required around and between those stairways.

Jim Dwyer and Kevin Flynn describe the devastating consequences of the new code.

> While the foundation of the trade center was being wrapped in a stupendous girdle of concrete, to stand against time and tide—it was three feet thick, more than half a mile long, and seventy feet deep—the stairways in the sky would be clad in a few inches of lightweight drywall. These stairways, bunched together, built for only a few hundred people at a time to walk three or four stories, would now have to carry out of the buildings the 12,000 people beneath the airplane impacts on September 11, 2001.[11]

What had been considered unacceptably dangerous when the World Trade Center was first being envisioned had, by the time the designs were complete, been written into law as the new standard of safety.

But as we have noted, the Twin Towers grew up at a time of changing paradigms of safety, and while stairway codes were being eased, a new threat was emerging from what had been considered the state of the art in safety. When construction began on the towers, standard practice in the industry called for spray-on fire protection containing asbestos. But soon, the EPA issued new standards for such fireproofing due to new evidence of the cancer-inducing effects of asbestos. WTC historian Gillespie notes that "construction had already reached the thirty-fourth floor" when "a difficult decision was made to stop using fireproofing with asbestos. . . . Then a difficult decision was made to change all of the spray in the World Trade Center, at the huge cost of $300,000. However, in retrospect, the decision to shift to the new material was the right one. To remove asbestos from just one floor of the center now [in 1999] would cost more than a million dollars."[12] In the span of just a few years, a standard, commonly used practice had gone from being considered the state of the art in safety assurance to being considered a health hazard and too dangerous to use.

In the abstract, safety is a laudable and worthy goal. It is in the realities of practice that safety becomes challenging. Given that even well-considered building codes and EPA standards—both explicitly meant to ensure safety—can and do change over time, what is a well-meaning engineer to do?

We can, in retrospect, be grateful that the WTC engineers went above and beyond what was expected of them. We can argue that they were surprisingly prescient in considering the effect an airliner would have were it to strike the towers. But just as the WTC building team needed to adjust to the new EPA standards, today's engineers must now adapt to the new knowledge and expectations of a post–September 11 world. Indeed, we are always adapting and updating our notions of what "safe" means and how to achieve it.

We Are Not Alone

No matter how carefully Yamasaki, Robertson, and their colleagues designed the World Trade Center and no matter how much they focused on safety, the outcome on 9/11 depended on the interaction of the towers' structure with the terrorists' actions. WTC 1 (the

North Tower) and WTC 2 (the South Tower) had roughly the same design but performed differently. The plane that struck WTC 1 hit the building higher than the plane that struck WTC 2, at less of an angle, at lower speed, and more centered on the face of the tower. WTC 2, by comparison, sustained far more serious damage to its corner support, and at a lower level (i.e., with far more mass above the damage), which led to Tower 2 collapsing first, though it was hit second. In addition, the damage to Tower 1 occurred on a floor where the stairwells were more closely spaced than anywhere else in the building, whereas in Tower 2, the stairwells at the site of impact were more widely scattered than elsewhere. As a result, the impact in WTC 1 destroyed all the stairwells at once, while in WTC 2, one stairwell remained outside the impact zone. The performance of the towers, then, and the resulting options for the towers' occupants, were clearly shaped not only by the structures themselves, but also by where the terrorists aimed the planes.

The outcome of 9/11 was further shaped by two events unrelated to the towers or the terrorists. Tuesday, September 11, 2001, happened to be both the first day of school for many students in the city and an election day. The result was that at 8:46 a.m., when the first plane hit Tower 1, only a fraction of the usual occupants had arrived at the building; the lower occupancy increased the speed of the evacuation and reduced the number of deaths.[13]

However careful the towers' designers might have been, they could hardly have been expected to know just where the terrorists would strike or just how many people would be in the buildings on their final day. The final caution about public safety, then, is that it never rests entirely in a designer's hands. Indeed, from the first days, the designers were part of a system that extended far beyond their influence.

The architects and engineers who brought the World Trade Center to reality faced any number of challenging dilemmas as they sought to ensure the safety of the towers, but chief among the challenges was the fact that the engineers and architects did not build the towers alone. They relied on the owners and developers to initiate the project, politicians and public agencies to make it possible, contractors to carry it out, and inhabitants to bring it to life. Though the engineers' role was critical, each of these players affected the safety of the building. To say this does not absolve engineers of their portion of responsibility for the creation of the towers as they existed; rather, it highlights the fact that engineers do not and cannot work in a vacuum. As a result, even if public safety could be paramount for engineers, it would not necessarily be so for the many others involved in the design process, making it all the more important for engineers to strive for safety even as they recognize it as unachievable.

Moving On

In January 2002, columnist Richard Weingardt called on his fellow engineers to "come forward to look after our nation's structures and stop them from being used as weapons for slaughter." But there will always be limits to what engineers can do to ensure safety. In response to an article on proposed post-9/11 changes to the New York City building code, Jon Magnusson (a member of the FEMA/ASCE study team) responded angrily: "There *are* changes that can be made in the construction of buildings that would increase toughness and the survival of occupants during certain types of attacks. However, when proponents claim that they can 'guarantee safety' during attacks similar to those of September 11, 2001, they are either ignorant or are intentionally misleading the public into a false sense of security." John Seabrook similarly pointed out in a poignant *New Yorker* article just two months after the attacks, "Engineers can't be asked to make every building safe from every possible event, yet that is just what people expect, and the engineers try to meet these impossible expectations." Seabrook and others have pointed out that perfectly safe structures would not only be prohibitively expensive, but disturbingly fortress-like as well.[14] But if engineering is design under constraint, then one cannot stop designing simply because constraints arise. Still, guidance for moving on would be helpful.

Every engineering society in the United States has a code of ethics to guide the conduct of its members. In the mid-1970s, the majority of these societies revised their codes. The most notable revision was the inclusion of a new canon, usually placed at the start of the code. One common version enjoins engineers to hold the public health, safety and welfare paramount. In many ways, these codes do not reflect reality. They are intended as guidelines or ideals toward which to strive, not as rules or standards to be followed (in the sense of a building "code," for instance). Put differently, it's all well and good to say that public safety is the most important consideration, but how does that play out in the day-to-day work of an engineer, particularly when the pressures of time, money, and expediency impose themselves? Although ideals are worth striving for, actual events force the interpretation of canons such as the one regarding public safety. What does it mean in practice for public safety to be paramount?

The WTC was perhaps more complex than many engineering projects, but it provides excellent examples of the tradeoffs that are inherent in engineering practice. Engineering is never about placing one specification above all others. Does this mean that codes are pointless or that public safety should be removed from them? Certainly not. Engineers strive for safety. This has been true for all of engineering's history. The issue in question is not whether engineers value safety, but rather how safety is defined, what its limits are, and who is responsible for ensuring it. The answers lie as much in the context as in the technology of design.

What will not change over time is humans' essential role in engineering; technology, by definition, cannot exist without human intervention. Both the cause and prevention of technological failures must therefore rest in human hands as well. In emphasizing that engineers are human, Henry Petroski suggested that the best engineers are humble engineers—sure of their abilities but willing to learn.[15] Ironically, many prospective engineers are drawn to the field because it seems more objective and certain and clear-cut than the humanities, and much of the coursework in undergraduate engineering reinforces this perception. In truth, engineering design is as subjective and uncertain an undertaking as one might hope to find.

FOR FURTHER EXPLORATION

1. Consult an engineering design textbook to learn how to develop a design matrix (also known as a decision matrix or rubric). In what ways can such a structured approach to design help engineers make appropriate decisions about safety and other factors? What limits does this approach present?
2. Select two automobile models—one contemporary and one from a generation or more ago. Compare the safety features on each. What is standard equipment now that was either optional or unavailable in the earlier vehicle? Choose one change to explore in more detail—how did it come about? Was it legislated? Did the change follow a notable accident or a new technological discovery? Did one car company lead the way? Did the change require any tradeoffs? What might such developments suggest about future safety improvements?
3. Select an ordinary household object (something as simple as a pair of scissors or as complicated as your microwave oven) and brainstorm as many potential failure modes as you can think of, remembering that a design need not be used as it was intended. Can you identify ways in which the designers meant for such failures to result in as little harm as possible? Can you think of additional protections that could or should have been included?

CHAPTER FIVE

"Architectural Terrorism"

Why Moderation Matters

I think there are times when logic just isn't the right way to think.

LESLIE E. ROBERTSON

Engineers must be society-wise as well as technology-wise.

WARREN J. VIESSMAN, JR.

Consider a pair of questions routinely asked by Professor Siva Vaidhyanathan at the start of a course on globalization that gained new meaning in the wake of September 11, 2001. Vaidhyanathan routinely began the class by asking students, "What would you die for?" and "What would you kill for?" When asked in the days before 9/11, the questions were, according to Vaidhyanathan, "merely academic." Within a week, however, the full meaning and power of those questions had hit home. We all—terrorists and patriotic Americans alike—have our limits, a point past which we will not go, even if our standing firm will result in death.[1]

Professor Vaidhyanathan used these questions to help students understand that no matter how foreign some ways of thinking may at first appear to us, part of the process we call critical thinking requires digging deeper, in an effort to make sense of what may at first appear to us to be nonsensical. We will understand the world more thoroughly if we avoid dismissing perspectives, ideas, and actions that are unfamiliar or uncomfortable, and instead insist on asking why a rational person would believe or do what seems strange to us. What experiences have led them to where they are, and how do they differ from our own? That skill of perceiving the world through others' eyes is essential to engineering practice, for the structures, products, and procedures that engineers produce are meant to be used (and may at times be misused) by others.

It is, of course, important to note that making sense of another's way of thinking does not require that we agree with that thinking or condone the behavior that results from it. As Aristotle famously noted, "It is the mark of an educated mind to be able to entertain a thought without accepting it." What Professor Vaidhyanathan's questions highlight so brilliantly is that though we do not all think alike, there are some underlying, fundamental commonalities. We are each driven by our view of the world, we each have passionate commitment to something—some value, some person, some belief—and under the right circumstances, we are each propelled to act.

Technology and Terrorism

In the tense and confusing months following 9/11, as Professor Vaidhyanathan was encouraging students to search for common ground with the terrorists (not to excuse, but to understand them), one New Yorker thought common ground was all too easy to find. Eric Darton had long been a critic of the excesses he saw represented by the World Trade Center, but writing in the aftermath of the 1993 bombing of the World Trade Center he took an astonishing stance, arguing that in some meaningful ways, the designers of the Twin Towers were themselves no better than terrorists, that there was "a kindred spirit linking the apparently polar realms of skyscraper terrorist and skyscraper builder."[2]

After 9/11, he again suggested that in fact the two professions of architecture and terrorism were but mirror images of each other, two sides of the same coin: "To attempt creation or destruction on such an immense scale requires both bombers and master builders to view living processes in general, and social life in particular, with a high degree of abstraction. Both must undertake a radical distancing of themselves from the flesh-and-blood experience of mundane existence 'on the ground.'" Darton went on to argue that the similarities do not always end with the approach, but may also appear in the results of both terrorists' and architects' work. "How far does a structure have to veer from fundamental considerations of human life and safety," Darton asked, "before we can recognize it as a manifestation of terrorism?"[3]

At first glance, Darton's claim strikes the reader as inappropriate, if not offensive. Darton himself knew he had "climbed out on the end of a very long limb where I could expect precious little company." Surely terrorists, whose aim is to destroy life and property—and thus security—could not be more different from engineers and architects, whose very purpose is to benefit humankind by designing structures that shelter life and property, and thus provide security. Indeed, Darton addressed head-on a topic that few others had been willing to address, even indirectly. Many writers had, in those early months, taken it as a given that the terrorists were to blame for the devastation of 9/11, and as the *New York Times* noted in 2002, until nearly a year after the collapse, few were willing to ask if the structure of the towers themselves (and thus those who had designed that structure) could have played a role. Those who did broach the subject tended to lionize the designers and vilify the terrorists, setting up a neat dichotomy of good versus evil. These authors suggested that perhaps the terrorists had won this battle, but the goodness of technology would win the war.[4]

By contrast, Darton dove in headlong and dared not only to ask if the designers might have played an inadvertent role, but also to suggest that the designers were in many ways the equivalent of terrorists. Designers and terrorists, instead of representing good and evil, respectively, were two faces of the same creature. The Roman god Janus, to whom Darton referred in the title of his essay, is generally depicted with two faces, one looking to the past, the other to the future, or one looking to war, the other to peace. He is the god of doors and beginnings and as such symbolizes potential, for both good and bad. Darton chose this imagery intentionally, suggesting not that engineers or architects and terrorists are equally evil, but that both draw on a single set of skills and viewpoints, with the potential for good and evil, depending on the uses to which they are put.

Darton's point is that designers and terrorists both rely on abstraction to achieve their goals: at the most critical stages, they must remove people from their equations in order to accomplish their feats of creation and destruction. In an early, online version of his essay, Darton put it starkly: "Package fifty thousand people in a ten million square foot office block accounting for weight and windloads and, as Yamasaki did, proclaim it a 'symbol of world peace.' Sure, no problem. And on the other end: calculate the structural properties of the target, the projectile's velocity on impact, the necessary payload of jet fuel. No problem. You just do the mathematics."[5]

Designers, though they are clearly building skyscrapers for human use, at key points lose themselves in detailed mathematical analyses from which humans are necessarily absent. The equations and forces that determine if a building will stand or fall do not rely on the vagaries of human behavior (and therein lies their beauty, as far as some are concerned). Similarly, terrorists, though they are clearly intending for deaths to occur, also lose themselves in abstraction, ignoring the human tragedy that will surely result from their actions in order to achieve the greater good (however they happen to define that—as a joyful afterlife, as a world in which their religion reigns, etc.). In Darton's words, "For the terrorist and the skyscraper builder alike, day-to-day existence shrinks to insignificance—reality distills itself to the instrumental use of physical forces in service of an abstract goal. Engulfed by their daydream, they are 'no longer aware of the outside universe.'" Darton notes that lead hijacker Mohammad Atta was in fact trained in architecture, and that his skills in planning and analysis helped ensure the success of his endeavor just as WTC architect Minoru Yamasaki's similar skills helped ensure the success of his efforts. Abstraction, thus, was necessary for both the creation and the destruction of the Twin Towers.[6]

"Existential pleasures" is the term civil engineer Samuel Florman chose to describe the joy to be found in such abstraction: "Every engineer has experienced the comfort that comes with total absorption in a mechanical environment. The world becomes reduced and manageable, controlled and unchaotic. For a period of time, personal concerns, particularly petty concerns, are forgotten, as the mind becomes enchanted with the patterns of an orderly and circumscribed scene." But as Florman notes on the very next page, "It is crucial that the engineer remember that the 'forgetting about life' is only for certain select periods of time, and must not be allowed to become a permanent condition. The joy of engrossment in the mechanical, like all of the existential pleasures of life, has the potentiality of becoming a destructive obsession."[7]

To their designers, the Twin Towers, like the first nuclear weapons and countless other complex technological problems, became at a certain point in the design process a puzzle to be solved. Richard Feynman, who had worked on the Manhattan Project, described the nature of his involvement in developing the atomic bomb: "You see, what happened to me—what happened to the rest of us—is we *started* for a good reason, then you're working very hard to accomplish something and it's a pleasure, it's excitement. And you stop thinking, you know; you just *stop*." For a group of scientists who had been raised on the notion of "pure science" and particularly for those who, like J. Robert Oppenheimer, declared themselves and their science as operating outside of petty everyday concerns, it became all too easy to become subsumed in the details, leaving questions of purpose and meaning and ethics to others.[8]

Darton uses the term "Janus Face" to suggest that abstraction, problem solving, and attention to detail are in themselves neither good nor evil skills. It is the uses to which those skills are put that we must approach with caution. Though Darton is willing to align terrorists squarely on the side of evil, he is not prepared to equate architects and engineers with good. Designers' intentions may be focused on creation rather than destruction, but creation can be evil and destruction can be good. If creation produces buildings that entomb their occupants (figuratively or literally) or simply dehumanize them, it surely is evil. If destruction removes an unhealthy or unsafe structure, paving the way for something better, it surely is good. Thus, Darton determines that although their damage may be more subtle, designers can and do inflict as much harm as terrorists.[9]

What is it then that separates them from each other? Terrorists are, in general, both highly educated and uncommonly dedicated to their community and their cause. But the same might be said of most engineers. Consider, for example, the claim made by structural engineering columnist Richard Weingardt in discussing how his profession ought to respond to 9/11: "American engineering expertise helped bring the country to where it is; now's not the time to abandon it." The National Academy of Engineering, in announcing its Grand Challenges for Engineering project, confidently asserted that "throughout human history, engineering has driven the advance of civilization" and expressed assurance that the future would depend on the contributions of engineers: "In each of these broad realms of human concern—sustainability, health, vulnerability, and joy of living—specific grand challenges await engineering solutions." Engineers are not alone in this conviction, as is made clear by the likes of *New York Times* columnist Thomas Friedman in his popular book *Longitudes and Attitudes,* in which he suggests that increased technological education would improve the world.[10]

These claims seem innocuous enough, yet at heart they reflect a deep, even unquestioning, faith in technology's benefits. It is a faith that infuses our experience with skyscrapers more specifically, as *New York Times* journalists Dwyer and Flynn explain: "For decades, the generations that rode higher and higher into the upper floors of skyscrapers had taken it on faith that the evolution of such buildings had been solely a story of progress, of innovations and enhancements that made new buildings safer than the old."[11]

But in the weeks after 9/11 New York City rabbi Brad Hirschfield astutely noted the danger of excessive faith: "It's so easy to get wrapped up in a messianic vision of how the world could be." In his view, "religion drove those planes into that building." Religion has the power to stir people to do remarkable things, for better and for worse. A similar lesson might be learned about technology, which both enabled the buildings to rise from the bedrock and brought them back to earth. If religious fervor has the potential to lead to terrorism, what might come from technological fervor?[12]

Lest we be tempted to dismiss the terrorists as delusional, it is worth considering where to draw the line between fervor and fantasy. Studies of terrorists make it clear that these individuals are neither uneducated nor mentally unstable. Rather, they are often highly educated and quite sane.[13] (Several of the 19 hijackers held college or graduate degrees, and all had been living what were in many ways "normal" lives.) One characteristic that distinguishes them from other educated individuals is the nature of their education, as indicated in the 9/11 Commission report—highly technical and with little in the way of humanities and social sciences:

> Millions, pursuing secular as well as religious studies, were products of educational systems that generally devoted little if any attention to the rest of the world's thought, history, and culture. The secular education reflected a strong cultural preference for technical fields over the humanities and social sciences. Many of these young men, even if able to study abroad, lacked the perspective and skills needed to understand a different culture.[14]

These words could just as easily be used to describe the education of many American engineering undergraduates.

Sociologists Diego Gambetta and Steffen Hertog write in "Engineers of Jihad" that "graduates from subjects such as science, engineering, and medicine are strongly overrepresented among Islamist movements in the Muslim world." In their study of over 400 members of violent Islamist groups, nearly 70 percent had higher education (compared to under 25 percent for the populations from which these members came), and roughly 45 percent of those with degrees had studied engineering—by far the most common degree of those in the study.[15]

According to Gambetta and Hertog, plausible arguments exist that terrorists are both drawn to and inspired by technical education, but that whatever the causal link, "a disproportionate share of engineers seems to have a mindset that inclines them to entertain the quintessential right-wing features of 'monism'—'why argue when there is *one* best solution'—and of 'simplism'—'if only people were rational, remedies would be simple.'" Engineers who wish to distinguish themselves from terrorists would do well to emphasize a broad approach to problem solving.[16]

The causal arguments may be controversial, but the terrorists' technical education supports the idea that terrorists are far from irrational, despite the fact that their beliefs and tactics may seem outlandish. Jon Krakauer, in his excellent book *Under the Banner of Heaven*, described the challenge inherent in "distinguish[ing] an unorthodox or bizarre faith from delusion." The ideas of the book's central figure, Ron Lafferty—a Mormon fundamentalist accused of murdering his sister-in-law and her baby "in order that [his evangelical] work might go forward"—seem "strange," not because they're irrational, but because "they are so uniquely his own." The courts struggled to determine whether Lafferty was mentally competent to stand trial. Psychiatric experts called to testify disagreed as to whether he suffered from a personality disorder or whether he was merely a religious zealot, defined as "someone who has an extreme, fervently held belief and is willing to go to great lengths to impose those beliefs, act on those beliefs." In the end, the courts determined that Lafferty's ideas, though odd and dangerous were "coherent" within themselves and therefore "rational" in the strict sense of that word.[17]

I include Lafferty's story not to suggest that we make a value judgment about one faith versus another, for part of Krakauer's argument is that such assessments are impossible to make in any objective way. The point for our purposes is rather that a fervently held faith needn't be irrational for it to lead to dreadful consequences. Perhaps the lesson to draw from September 11 does not have to do with improved fireproofing or stronger floor trusses at all, but is instead the realization that the largest danger in engineering lies in unquestioned and fervent faith in its righteousness, to the exclusion of other viewpoints.

Moderation through Diversity

What separates engineers and terrorists? Such a question is the equivalent of asking for a definition of engineering. Engineering as "design under constraint" is an elegant but inadequate description, for it addresses neither the participants in nor the purpose of the enterprise. Engineering is not distinguished from terrorism by its reliance on the scientific method and the laws of nature, for terrorists, too, make use of these. Nor is engineering set apart by having the "public interest" or "public benefit" as a goal, for terrorists, too, claim these goals. Not even a faith in technology is adequate to distinguish engineers, for again, terrorists can believe in the power of technology as a means to their desired ends.

The answer, I would suggest, is that alongside a reliance on science and a conservatism that protects the public, there should exist in engineering an openness to a diversity of viewpoints and a belief that such diversity will lead to the best possible result or design solution. Bill Wulf, past president of the National Academy of Engineering, has argued that creativity is essential to design and is best achieved by a diverse team willing to challenge and improve on one another's ideas. This openness to diversity, then, both extends and moderates the results of engineering work.

This ideal of engineering is not, however, fully reflected in reality. The decreasing percentage of women and underrepresented minorities entering the field, the concern in professional societies with the increase of non-U.S. engineers, and curricula that continue to view liberal studies as of secondary importance all suggest that diversity has yet to take center stage in the education or practice of American engineers.

The percentage of women among American undergraduate engineering students has been decreasing since 1999, when it reached a peak of 19.8 percent; the percentage in 2007 stood at 17.5, the lowest since 1993. Undergraduate enrollment of African American students in the United States has also been decreasing since the 1999 high of 7 percent, to under 6 percent in 2007. Native American enrollment in undergraduate engineering programs slipped between 1999 and 2000 from 0.7 percent to 0.65, where it has remained ever since. The picture for Hispanic Americans is less steady, with the percentage bouncing up and down in those years from a low of 7.5 (in 2007) to a high of 8.4. Asian American enrollment, too, has not followed a strong trend, varying between 11 percent and 11.8 percent in that same eight-year period. Explanations and interventions abound for these demographic patterns, but as a profession we have not yet fully accepted the notion that engineering is shaped by those who practice it.[18]

Globalization has been a buzzword in engineering education for a decade, and yet for every article on the importance of preparing American students to work internationally, there is an article about America "losing ground" to other countries, particularly China and India. Rather than viewing the increase in technological education overseas as providing potential for new partnerships or collaborations, there has been an unfortunate tendency to view non-U.S. engineers as competition.[19]

In discussing both domestic and international diversity, we must beware of excessive conviction in the status quo and in the immutable nature of engineering itself, for in truth it is a field shaped by its practitioners and the times in which they live. The ASCE has been actively engaged for several years in exploring "the changing civil engineering landscape" and in offering recommendations concerning the "body of knowledge" necessary for tomorrow's civil engineers. According to a recent report by the ASCE Body of Knowledge Committee,

> Civil engineering must restructure its 150-year-old educational model to meet the challenges of the 21st century. . . . Today's world is fundamentally changing the way civil engineering is practiced. Complexity arises in every aspect of projects, from pre-project planning with varied stakeholders to building with minimum environmental and community disturbance. The 2001 ASCE report *Engineering the Future of Civil Engineering* (www.asce.org/raisethebar) highlights the significant and rapid changes confronting the profession, while recent events have demonstrated our vulnerability to human-made hazards and disasters. The risks and challenges to public safety, health, and welfare will continue to escalate in complexity, and the civil engineering profession must respond proactively.[20]

The events of September 11 and the turn of the millennium inspired the National Academy of Engineering as well. In 2002 alone, the NAE published a special volume of *The Bridge* devoted to engineering and homeland defense, and another focusing on engineering ethics. In addition, NAE's "Engineer of 2020" project, "a two-year vision-casting initiative on engineering in the future and educating engineers to meet the needs of the new era," will shape engineering practice for the next generation.[21]

Similarly, in the late 1990s, the Accreditation Board for Engineering and Technology introduced Engineering Criteria 2000, a complete overhaul of the criteria and the processes used to accredit undergraduate engineering programs in the United States. The new criteria, which have now been in place for nearly a decade, require engineering programs to prove that their graduates have the following knowledge and skills:

(a) an ability to apply knowledge of mathematics, science, and engineering

(b) an ability to design and conduct experiments, as well as to analyze and interpret data

(c) an ability to design a system, component, or process to meet desired needs, within realistic constraints such as economic, environmental, social, political, ethical, health and safety, manufacturability, and sustainability

(d) an ability to function on multi-disciplinary teams

(e) an ability to identify, formulate, and solve engineering problems

(f) an understanding of professional and ethical responsibility

(g) an ability to communicate effectively

(h) the broad education necessary to understand the impact of engineering solutions in a global, economic, environmental, and societal context

(i) a recognition of the need for, and an ability to engage in life-long learning

(j) a knowledge of contemporary issues

(k) an ability to use the techniques, skills, and modern engineering tools necessary for engineering practice.[22]

This list of educational outcomes calls for engineering graduates to have mastered not only the technical components of engineering, but also additional knowledge and skills that might be termed the "professional," or humanities and social science, components of engineering.

In response to ABET's requirement that engineering education prepare graduates in this broad range of areas, the Liberal Education Division (LED) of the American Society for Engineering Education produced a set of "Recommendations for Liberal Education in Engineering." A task force of LED members furthered these efforts by hosting a National Science Foundation–supported workshop at the University of Virginia in 2002 to continue the discussion of liberal education's role in engineering education. The task force described the importance of liberal education as a complex component of engineering education that touches students in a number of ways. At one level, it provides crucial knowledge and skills that are essential to the responsible and effective practice of engineering. At another level, it helps develop understandings that round out and complement engineering design and decision-making. Finally, it enriches the capacity to articulate and reflect on personal and professional values and contributes to the development of leadership ability.[23]

How different the profession might look if we took the ASCE Body of Knowledge Committee, ABET, and the LED committee at their word and considered humanities and social science perspectives and contributions every bit as important to good engineering as soil mechanics, thermodynamics, or differential equations.

Hope exists—in the form of groups like Engineers Without Borders, in the recent updating of the ASCE code of ethics to include sustainability as a goal, in the increasing popularity of service learning as a pedagogical tool, and in the development of new programs at Smith College and Olin College that emphasize engineering as a liberal art that requires integration of knowledge. The National Academy of Engineering, under the leadership of Bill Wulf, encouraged and assisted the engineering profession in examining and improving engineering education in the early twenty-first century.

What these initiatives have in common is a conviction that people must take center stage among engineers' concerns. Abstraction may help solve a tricky equation, but the true benefits of engineering come not from the engineered product per se, but rather from that product's ability to make people's lives better. Doing so requires an engagement with those people one wishes to help, rather than behaving as technological missionaries. Equally important in these initiatives, then, is the idea that engineers can offer only one piece of the solution—to truly serve the public, engineers must collaborate with that public. Here too is a reason why diversity matters in engineering—the more engineers reflect the population they serve, the more likely such collaboration is to occur.

Eric Darton ends his essay on architectural terrorism with this cautionary note: "We humans are born creatures of the earth and air, capable of functioning with our heads in the clouds—so long as our feet remain on the ground. Rising toward the stratosphere, though, we feel we have broken free of gravity. When that narcotic sense of weightlessness possesses us, it is not long before our ascent finds its opposite number in the terror of the fall."[24] As Professor Vaidhyanathan exhorts students, we should consider what factors might lead a rational person to make such a bold claim as the equation of architecture and terrorism. Whether we agree with his conclusion or not, Darton's reminder to keep our feet on the ground even as we dream in the clouds, like Samuel Florman's caution against "destructive obsession" with abstraction and isolation, reflects the fundamental need for balance in the engineering enterprise—balance of perspectives, of practitioners, of purposes.

FOR FURTHER EXPLORATION

1. How would you answer Professor Vaidhyanathan's questions from the start of this chapter?
2. Locate a description of engineering from each of three sources:
   - an undergraduate program website,
   - a middle or high school recruitment brochure (try checking a professional engineering society's website for its "outreach" efforts), and
   - an introduction to engineering textbook.

   How do they compare? What reasons does each give for encouraging participation in engineering? How do they compare engineering with other fields? What types of people is each likely to attract (or deter)?

3a. In a small group, list as many life experiences as possible that everyone in the group has in common. Were there any surprises on the list? Is the number of commonalities larger or smaller than you expected? In what ways might the things on this list help your group work well together? In what ways might you be hampered by them?

3b. In your same small group, list as many unique life experiences as you can identify for each person (i.e., what has each person done or experienced that no one else has done?). Identify at least one unique experience per group member. What surprises did you find? Was it easier or harder to come up with the list than you expected? In what ways might this diversity help your group function well as a team?

3c. What steps would your group need to take to benefit from both its similarities and its differences?

CHAPTER SIX

"These Material Things"

Passion and Power in Engineering

It just isn't possible for me to take the posture that the towers were only buildings . . . that these material things are not worthy of grieving.

LESLIE E. ROBERTSON

The 16-acre bathtub that once served as the foundation of the World Trade Center became on September 11, 2001, a vessel for enormous grief. The most searing form was that felt by family and friends who lost loved ones in the tragedy. Those of us fortunate enough not to have known victims personally were pained by the loss of so many fellow human beings.

A moment's reflection reminds us that there was grief too for intangibles. For our lost innocence. For our lost sense of security. For the life we will never again live. For reputations marred by affiliation with the disaster. For livelihoods lost amidst the rubble.

For those of us with a connection to the city, there was grief as well for the towers themselves. Vilified though they were at their creation, they had become familiar parts of the skyline, and their familiarity bred fondness. Rutgers University professor Angus Kress Gillespie, who had finished his book on the life of the Twin Towers just months before their death, reported in October 2001 feeling "like a kid with his two front teeth knocked out." Apart from the pain of the loss was the surreal sense that a part of him was gone. Those two incisors, so central to the face of New York City, were suddenly gone, leaving us staring at the gaping hole and seeing the surroundings in a new light.[1]

A few felt that physical loss more keenly. Leslie Robertson described the towers as his children, not twins at all, but two unique siblings with a proud and protective father.[2] As WTC structural engineer, he knew them as intimately as anyone, having done so much to bring them into the world. His fondness for them went beyond that of most New Yorkers. In one of the most heart-rending descriptions by an engineer of a disaster, Robertson reflected on the thoughts that haunted him in late 2001. After detailing "thoughts of the thousands who lost their lives as my structures crashed down upon them," Robertson confessed that he grieved for more than the people who died. "It just isn't possible for me to take the posture that the towers were only buildings . . . that these material things are not worthy of grieving." Robertson, the man whose reputation had been built with the Twin Towers, wistfully told an interviewer in the weeks after the attack, "That's how people introduced me"—with the towers. A part of his identity had collapsed with them. The reporter described feeling "the absence of the buildings" in Robertson.[3]

Leslie Robertson was not alone in grieving for the towers. Frank Lombardi, chief engineer with the Port Authority, was described by one journalist after the collapse as having "a manner of talking about the towers as if they had once been alive. He had obviously been thinking heavily about their deaths." Abolhassan Astaneh-Asl, an engineering professor at the University of California–Berkeley who'd come early to New York to study the collapse, bemoaned how personal the attack felt to many structural engineers: "It's your product, and they used it to kill people." The towers had been struck, but the blow resonated far beyond the 16-acre WTC site.[4]

Humanizing Our Designs

Admitting that the towers themselves are a source of grief need not diminish the lives that were lost. Grieving for the structural victims as well as the human victims is a sign of both the passion felt for our creations and the power that structures have to shape our lives in ways large and small. This passion and power are as much a part of the "material things" around us as the steel and concrete, plumbing and wiring of which they're constructed. Civil engineer and author Samuel Florman was thinking along the same lines when he wrote of the "existential pleasures of engineering"—that fundamental joy that accompanies creation: "The engineer's first instinctive feeling about the machine is likely to be a flush of pride. . . . The primordial joy of the successful hunt or the abundant harvest has its modern counterpart in the exhilaration of the man who has invented or produced a successful machine."[5] Understanding the dual nature of the grief following September 11, then, can tell us something about the nature of engineering practice itself and about the relationship of engineers to the structures, products, and technologies they create.

The towers were hardly the first or only buildings to be imbued with human characteristics. Matthys Levy and Mario Salvadori in their book *Why Buildings Fall Down* employ the human body as a metaphor for all buildings.

> A building is conceived when designed, born when built, alive while standing, dead from old age or an unexpected accident. It breathes through the mouth of its windows and the lungs of its air-conditioning system. It circulates fluids through the veins and arteries of its pipes and sends messages to all parts of its body through the nervous system of its electric wires. A building reacts to changes in its outer or inner conditions through its brain of feedback systems, is protected by the skin of its façade, supported by the skeleton of its columns, beams, and slabs, and rests on the feet of its foundations. Like most human bodies, most buildings have full lives, and then they die.[6]

Nor is it unusual for this personification to extend to the death of buildings as well. The History Channel produced a video about the buildings called *World Trade Center: In Memoriam*, consciously anthropomorphizing the towers. Princeton architecture professor M. Christine Boyer referred to the towers' *New York Times* "obituary" as though they had been famous citizens struck down unexpectedly, and whose death warranted similar coverage. Moustafa Bayoumi, a professor of English at City University of New York, described New York City after 9/11 as being "in mourning, with its gaping wound right there on the skin of Lower Manhattan."[7]

Mark Wigley, professor of architecture at Columbia University, explained these bodily metaphors:

> Damaged buildings represent damaged bodies. And it is the representation that counts. Terrorism is not about killing people, but about dispersing the threat of death by producing frightening images. Particular sites are targeted to produce a general unease. . . . Again and again, the towers are described with the same terms used for suffering people . . . the repeated use of expressions like "wounded buildings," "victimized buildings," "tortured structures," "death of the towers," and "death of the twins." In the grieving for those who died, there is also grieving for the buildings themselves.

Abolhassan Astaneh-Asl similarly invoked this parallel with living things when he described for a reporter the impending study of why the towers collapsed as "like doing an autopsy." "Forensic engineering" is hardly accidental, then, as a term for studying the death of these structures.[8]

Some buildings clearly possess such power that it is only through equating them with humans that we feel able to capture the role that they play in our lives. But the point here is twofold: first, that buildings are often thought of in human terms, and second, that the symbolism attached to buildings begins before their creation and lasts after their destruction. Mark Wigley again offers us some help understanding this symbolism and the power of our reactions to the destruction of an icon such as the World Trade Center:

> To begin to understand the depth and complexity of reaction, we need to go back to the simplest level. In the simplest terms, buildings are seen as a form of protection, an insulation from danger. . . . Furthermore, buildings are traditionally meant to last much longer than people. . . . To lose a building is to lose not simply an object that you have been living in or looking at but an object that has been watching over you. And when our witnesses disappear, something of the reality of our life goes with them. People are really grieving for themselves when they grieve for buildings.[9]

To understand fully the grief for the towers requires that we look past the integrity of their structure to the impact of their very existence.

What We Shape Shapes Us as Well

In 1986, Langdon Winner asked a question now famous among scholars in the discipline known as science and technology studies: Do artifacts have politics? In other words, do technological objects (i.e., "artifacts") hold in themselves the power to shape society? The suggestion that a bridge, for example, might stand for a particular political position, even long after its designer is dead and gone seems at first almost laughable. A bridge, after all, is designed to fit in a particular landscape, withstand certain specified loads, and enable the transport of people and materials from one location to another with greater ease than would be possible without it. It is, thus, merely a structure.[10]

And yet, consider that the placement and form of a bridge both allows and restricts passage. For each pair of locations joined by a bridge, many others are excluded (some even destroyed). The bridges of Manhattan, for instance—the George Washington, the Brooklyn, and others—each provide access to the island city, but each also has an endpoint at a specific location outside the city, benefiting some outlying towns and bypassing others. Each design decision about loads is tied to and reflects decisions about what type of traffic will and won't be allowed to cross—will pedestrians, trains, cars, and trucks all be given access, or will priority be given to certain modes of transportation? Consider, for example, the bridges of Venice, Italy, where water traffic reigns, or the skywalks of Minneapolis, Minnesota, designed to ease pedestrian traffic during the notoriously fierce winters of the northern Midwest.

For every form of traffic included or excluded from travel *on* a bridge, we must also consider the forms of transportation possible *under* that bridge. The Duluth Lift Bridge was designed around the shipping lane that runs underneath it (versus the light traffic above). The Duluth bridge itself provides access only to and from a small strip of sandbar across the harbor from Duluth proper and thus was designed around protecting the ship traffic below it. The Park Street railroad overpass in Madison, Wisconsin, was recently redesigned and rebuilt to improve safety for the pedestrians who must travel underneath it; the wider sidewalks and improved lighting and openness have made this a safer path to walk, particularly after dark. Robert Caro highlighted a final, powerful example of the importance of bridge traffic when he described developer Robert Moses' efforts to restrict lower-class residents of New York City from gaining access to Jones Beach by making the overpasses on the Long Island Expressway too low for public buses to travel underneath them.

Urban historian Raymond Mohl summarized the ways in which a series of technological developments in the late nineteenth century shaped the urban landscape and the lives of city residents. He began by noting how the skyscraper symbolized and allowed "corporate domination of the modern city" and went on to catalogue the variety of other technological influences on city life. As Mohl noted, suspension bridges, such as Roebling's Brooklyn Bridge, completed in 1883, expanded the territory of cities beyond their traditional river boundaries and "symbolized the indomitable and expansive spirit of the industrial era." Beginning in 1876, the telephone "eliminate[d] the necessity for the face-to-face contacts of the preindustrial period and brought a communications revolution to urban America." Likewise, the typewriter and cash register changed the way business was done within those new skyscrapers by increasing emphasis on efficiency. But, Mohl continues, "the technological innovations with the greatest impact in creating the new spatial order of urban America were those in transportation and building. Streetcars and subways, skyscrapers and tenements changed the texture of urban life and permanently altered the built environment of the American city."[11]

Transportation networks and building layouts literally shaped where people lived and worked and how they traveled between the two. Suburbs, for instance, exist only because bridges, cars, and mass transit make them possible. Penthouses are desirable only because of the elevators that make them convenient. These changes, begun in the late nineteenth century, took an extreme form in the financial district of Lower Manhattan in the years around 1970. The overwhelming majority of World Trade Center tenants did not live in the neighborhood where they worked (indeed, few residential buildings exist in the area); most of the 20,000 employees in this vertical city departed overnight. The most sought-after addresses in the Twin Towers were those closest to the top; why locate your business in one of the world's tallest buildings only to be on the fifth floor? Even more strangely, the most bustling region of the towers lay underground in the vast network of shops that lay outside the PATH train terminal; restaurants, bookstores, and newsstands all disappeared into this subterranean mall, where they found more traffic than on the formerly crowded streets above. When the Twin Towers opened in 1971, the most valuable real estate no longer lay on the streets of Radio Row (as the neighborhood had been known in the first half of the twentieth century), but instead could be found in the suburbs, the skies, and underground.

Technological artifacts do have the ability to reflect and reinforce certain values. As the examples above make clear, the notion that technology itself is neutral and is given value only by its users is untenable as a universal truth. No matter how badly New Yorkers may *want* to ride public buses to Jones Beach, they are incapable of doing so, so long as Moses' bridges stand. His political views became hardened in concrete and steel artifacts that force users to either abide by or work around his beliefs on the value of racial and class divides.

The World Trade Center, too, had specific uses built into it. Architect Minoru Yamasaki had designed the plaza and ground-level entryways to the towers to mimic the Piazza San Marco in Venice and meant them to provide a welcoming plan and a human scale to these behemoths. In practice, however, few tenants or visitors to the center approached the towers from this direction. In his 1999 book about the towers, Eric Darton described approaching the buildings: "Though it is possible to enter the trade towers by walking across the raised plateau of Austin Tobin Plaza, this requires a deliberate detour because the design of primary access routes works against the street-level approach. It is far more likely that we have navigated a warren of corridors, staircases, and escalators beneath the complex via the vast enclosed WTC Mall submerged beneath the plaza." For Darton, this miscalculation by Yamasaki provided evidence of the danger of architects' habits of abstraction; on paper, Yamasaki's plan for a welcoming plaza made sense, but it did not take into account how actual people would made use of the space once it was built, earning him Darton's moniker: Yama, Architect of Terror.[12]

Points of View

The World Trade Center also makes clear that whatever values or politics or patterns of use may be built into a structure, the way that structure is viewed depends upon who is doing the viewing. The terrorists and the towers' creators both saw the WTC as a symbol of power and pride, but the terrorists were offended rather than uplifted by this brazen declaration of superiority. These two groups represent two extreme points of view, but they were hardly alone in ascribing meaning to these structures.

The World Trade Center was, without question, a technological marvel; its framing, elevators, and motion studies all marked it as state of the art. The director of the Skyscraper Museum in New York City described the Twin Towers as a product of the same age as jumbo jets, both conceived and developed in an age of giantism and bold engineering, enabling each other. But the Twin Towers were from the start intended to do more than simply supply vast quantities of modern, flexible office space. They were envisioned as status symbols that would draw powerful tenants to an address made famous by its record-breaking height, though Manhattan did not need the 10 million square feet of space being offered. They were envisioned by their planners at the Port Authority as beacons to New Yorkers, advertising the new life being given to Lower Manhattan in the form of a "Financial District," replacing what had been a vibrant, if working-class, community. They were envisioned as exclamation points on the claim shouted worldwide that New York City still intended to be the center of world trade, in spite of evidence to the contrary—other ports were drawing more traffic, and electronic banking was reducing the need for a centralized marketplace. The iconic (as opposed to real estate) value of the towers led one commentator to suggest after 9/11 that "the question of how to replace nine million square feet of office space is irrelevant. If anything, the issue is how to replace the more than two million square feet of façade."[13]

The Twin Towers were not unique in having a symbolic value that outweighed their practicality. Architect Adrian Smith has noted that "there are only 15 to 20 tall buildings in the world that are considered 'icons.' They tend to be owned mostly by governments or government agencies, and they're done primarily for ego purposes, to create a new landmark for a city or a new symbol for a country. They're usually not built from market forces." His colleague Thomas Fridstein agreed that "the 100-story building as an office building, has long been impractical. None of these had been built to be practical from a real estate investment point of view." The Twin Towers may well have been cutting edge, but they were certainly not practical, as evidenced by the long struggle to find sufficient tenants to fill them to capacity. Even one Port Authority board member took to referring to the World Trade Center as "a giant white elephant."[14]

And yet the claims for the towers became bolder, and the symbolism became more widespread. Minoru Yamasaki, WTC architect, famously declared that his masterpiece not only symbolized world trade but was also "a living symbol of man's dedication to world peace."[15] Souvenir stands, postcards, and guidebooks to New York City seemed invariably to use the recognizable towers as stand-ins for all of Manhattan. Commuters and tourists alike relied on the towers as signposts marking their progress to and around the city.

The Twin Towers, brazen declarations of Manhattan as the world's focal point, did engender a certain fondness, if only because their insistent size quickly made them familiar. One memorialist noted, "The Twin Towers made their great impression by sheer arrogance." But the towers did not evoke positive feelings in all observers. Their earliest critics were those business owners along Radio Row like Oscar Nadel, whose livelihoods were to be sacrificed in the name of progress. Landlords elsewhere in the city were next in line to condemn the excess represented by the buildings. To these two groups, the towers symbolized the end of a way of life, and some would never recover, financially or emotionally.[16]

(It is worth noting as an aside that given the long history of rebuilding in New York City, Radio Row itself had hardly been an original tenant of this region of lower Manhattan. It had replaced the Syrian Quarter, which in turn had been built on an eighteenth-century African Burial Ground containing some 20,000 bodies, and so on back to the first Dutch farmer who had taken the ground from the Native Americans of Mannahatta.)[17]

The New York Skyline, without the Twin Towers. Much of the debris from the Twin Towers was taken by barge to a New Jersey landfill, where it could be sorted and either saved or recycled. ASCE/FEMA, *Building Performance Study*, www.fema.gov/rebuild/mat/wtcstudy.shtm.

As the Twin Towers were completed in 1971, architectural critics quickly decried them as well. The towers were out of proportion to the rest of the cityscape. They were boxy and unimaginative next to the graceful and elegant Chrysler and Empire State Buildings. (One joke even referred to the Twin Towers as resembling the shipping boxes for those two elegant buildings.) They were impossibly huge, dwarfing, even swallowing their tenants. As Lewis Mumford described them in his book *Pentagon of Power*, the towers were an "example of the purposeless giantism and technological exhibitionism that are now eviscerating the living tissue of every great city." When journalist John Hockenberry declared, "Everything that is best in America was embodied in these buildings," Marshall Berman responded that, in fact, "they were the most hated buildings in town." Much like the powerful millionaires who brought them into being, the towers were sterile, intimidating behemoths that overlooked, in both senses of the word, the city and its residents. They were, in short, above the city rather than of the city.[18]

The Twin Towers were indeed symbols, though not necessarily of freedom and peace. But as Yamasaki, Robertson, and countless architectural critics of the towers understood, once constructed, buildings take on a life (and a symbolism) of their own. Angus Kress Gillespie began his book on the World Trade Center with a question: "What do the Twin Towers mean?" It is a philosophical-sounding question, but not a rhetorical one, for Gillespie meant to answer it. He responded that the towers "might be taken to symbolize the Manhattan skyline, or the City of New York. . . . But on a higher plane, the Twin Towers might be taken to symbolize American exceptionalism, or American capitalism, or even America itself." He went on to explain: "If the World Trade Center was a symbol of America, then it was the perfect target for enemies of America." Indeed, the terrorists who bombed the towers in 1993 as well as those who attacked them in 2001 were drawn to them precisely because of that symbolism. If a massive death toll was what they were seeking, they could have found other, more efficient and productive targets. But for maximum symbolic effect, the towers were far better suited to the task. As one art historian noted in the *New York Times* just days after the collapse, "No one seems to have seen them as clearly as the men who destroyed them."[19]

And indeed, for many of us, it took the destruction of the towers to show us what they meant. In the troubling weeks after 9/11, as we each sought ways to contribute to the recovery (and as the first civil engineers were converging on Ground Zero), architect Richard Keating suggested that the role he and his colleagues might play would be "to portray the poetry of these tall buildings and what they mean to us. In my opinion, the attack on September 11th had to do with a symbol that was extraordinary in its imagery. It represents not only our country but our financial system." As Sharon Zukin described it, "Once we gazed upon this site as a landscape of power, but since September 11, we have viewed it in sorrow—as if it holds both the dark side of grandeur and our unspoken fears of decline." For her colleague David Harvey, the meaning of the towers and of the attack depended upon where one called home. In New York, Harvey argued, "September 11 was represented as a local disaster of horrific magnitude with unfathomable causes and unthinkably tragic personal and local implications. Nationally, the media immediately followed President Bush in construing it as an attack upon 'freedom,' 'American values,' and the 'American way of life.' " "There were, clearly," he added, "different local, national, and international ways of understanding events."[20]

Osama bin Laden himself, shortly after the attacks, proudly declared: "Those awesome symbolic towers that speak of liberty, human rights and humanity have been destroyed. They have gone up in smoke." It is with this in mind that the History Channel could refer to the towers' destruction as "a symbolic act more than a murderous act."[21]

Choices, Past and Present

And thus the answer to the questions of why the towers fell when they did and who was responsible and how we should build in the future lies not so much in discussions of physical reinforcements or battlements as in consideration of what we mean to say with our structures. In the weeks after the collapse of the towers, as the official death toll dropped improbably, Leslie Robertson was interviewed for a *New Yorker* article that asked noncommittally, "Why did the World Trade center buildings fall down when they did?" Robertson had been praised for his design, which had allowed perhaps 15,000 or 20,000 souls to evacuate the towers before their collapse. He privately responded to such praise with a wrenching thought: "But that I had done a bit more . . . Had the towers stood up for just one minute longer . . . It is hard."[22] One can ask, as Robertson did, what would have happened if the towers had stood one minute longer, but one can also wonder what would have been if the towers had never been built in the first place—if the vast sums of money and time spent to build them had been expended in other ways.

In the 1960s, Guy Tozzoli of the Port Authority rejected architect Minoru Yamasaki's original design for the World Trade Center, insisting that Lower Manhattan needed a full 10 million square feet of additional office space—even while other properties went vacant. It was not so much that the towers as originally proposed were too small for Tozzoli's liking, but rather that they were too modest. Tozzoli and his colleague Austin Tobin wanted the World Trade Center to be a commanding presence and the headquarters of world trade. But even as the towers were being planned, many of the businesses that might have filled the space were moving out of New York, for the business of trade itself was changing, making physical proximity less and less important. As the consulting firm that Tobin hired recommended, a World Trade Center "would have to be unusual in nature and spectacular in proportions to act as an irresistible magnet to such lukewarm prospective tenants." And yet, even at a record-breaking 110 stories each, the towers were far below capacity when they opened to tenants in 1970, and remained so for many of the years of their existence. Aaron Kress Gillespie has suggested that the reason for the towers' collapse lies in the reason for their creation.[23]

When after 9/11 the question began to be asked, How shall we build in the future? several prominent engineers argued that although sensible precautions ought to be taken, we should not go to the extreme of making building security the paramount concern of engineers. They recognized the symbolism of buildings and building styles. Christopher Foley, for instance, in his 2001 article "Why the Towers Fell," called on engineers not to succumb to pressure to build "bunker-type" buildings in response to the collapse of the Twin Towers, arguing that such an approach would be incommensurate with the very freedoms being attacked by the terrorists. "Where," he asked, "is freedom expressed in concrete bunkers or underground tunnels? Where would we be without tall buildings?

We would live in a very lifeless place, a place unworthy of being called the United States."[24]

It is poetic to equate tall buildings with freedom, but these two buildings were not just any skyscrapers. They resulted from and represented choices and attitudes and desires—capitalism, control, and ego. Decisions about the future of design must be shaped with an explicit and conscious recognition of the breadth of factors affecting the dynamics of design. Debates over what to do at Ground Zero after the cleanup were really but continuations of the decades-old dialogue about the meaning of the towers.

On one side were those who, like Foley, wanted to defy the terrorists, literally and figuratively, by rebuilding. Some proposed recreating the towers just as they had been. Others suggested erecting an even taller structure, to symbolize triumph over the attackers. On the opposing side of the debate were those who argued that rebuilding would be at best foolish, begging for a repeat attack, and at worst disrespectful, insulting those whose final resting place lay in the dust of the site. Today the site plans call for a compromise, with a "Freedom Tower"—clearly symbolic in its name as well as its 1,776-foot height—and a memorial plaza.

Moving On

Psychologists tell us that following tragedy it is natural to experience grief in stages beginning with shock and moving through bargaining, depression, and anger, to acceptance. Shock was widespread in the hours and days after 9/11: how could such a thing have happened? It seemed too unreal, too movie-like, especially when viewed on television, indistinguishable from a Hollywood disaster film. During the bargaining phase of grief, counselors say, some "may ruminate about what could have been done to prevent the loss"; surely the engineering profession experienced this stage in its investigations. For some, a form of depression left them feeling hopeless: what could we possibly do to prevent such attacks in the future? It all seemed so unpredictable, so random, so unpreventable. For the lucky, this depression gave way to anger—a sign of fighting back against that hopelessness. Alongside calls for vengeance, bold claims began to appear about defying the terrorists by rebuilding the towers, bigger and stronger than before.

The challenge now before us lies in acceptance. As one grief resource explains, "healing occurs when the loss becomes integrated into the individual's set of life experiences." For the engineering profession, acceptance and healing will come when the death of the World Trade Center becomes a part of us, when we recognize the factors that led to its creation and its destruction as equal parts of the engineering enterprise.[25]

Distance helps us to cope with grief by allowing us to view devastating events from a broader perspective. Samuel Florman described our tendency to adjust our views of the structures around us.

> Even in an era of disenchantment with technology, intimacy serves to enhance the attractiveness of the machine rather than diminish it. A patina of familiarity softens its surfaces and enhances its appeal. The prime example is what has happened to the locomotive. Once regarded as an ugly monster, its beauty and charm are now widely acclaimed.[26]

The passage of time led some to view the Twin Towers more charitably as well during their lifetime, but has not yet allowed us to adjust to their deaths. Consider a juxtaposition that appeared in *ASCE News* in May 2005. On one side of the page lay a column on the NIST report of the WTC collapse; abutting this news was a story declaring "Iron Truss Bridges Are Works of Art." The former was concerned with the technical minutiae of the towers' demise. The latter looked at nineteenth-century utilitarian artifacts that had survived into the twenty-first century and described the beauty and elegance in the metal framework of these objects. It is important to note, though, that the beauty of the bridges lay in the technical—to appreciate the grace and accomplishment embodied in these spans requires a grasp of the very dynamics being studied by the NIST team. Though we may require time to see the connections clearly, the aesthetics and the physics of engineering are inseparable.

What has any of this to do with preparing engineers for the work they will do? In the previous chapter I spoke about the crucial role of creativity in engineering. The truly imaginative engineer will see that although courses in subjects like statics, thermodynamics, and fluid mechanics may make it possible to do engineering design, courses in art history, English literature, or political science will help place engineering work in its broader context and provide a deeper understanding of and appreciation for the engineering enterprise and its impact. Art history can help explore the prevalence and power of symbolism and aesthetics in shaping as well as reflecting culture and values. Literature can demonstrate the ability of abstract ideas—ranging from love to war—to move and inspire individuals and civilizations. Political science can clarify the role of governments and public policy in directing the place and form of technology.

Courses in economics, foreign languages, anthropology, and other subjects can challenge the imaginative and introspective engineer to consider a variety of viewpoints and research styles, and ultimately to grasp this fundamental truth: there are a multitude of ways to understand and contribute to the world. A mature and professional engineer needn't enjoy each of them equally to value their existence and recognize where they might intersect with or contribute to the engineering approach to the world.

Symbolism was essential to the life and the death of the World Trade Center. Developer David Rockefeller and architect Minoru Yamasaki envisioned the towers as signifying world trade and world peace, respectively. The business owners in Radio Row whose livelihoods were destroyed to make way for the towers' creation and the terrorists whose lives ended in the towers' destruction saw them as symbols of the domination of American capitalism. The answer to why the towers fell requires an appreciation of the power of this symbolism.

In the early twentieth century, philosopher Martin Heidegger despaired of the assumptions being made about technology and its promise as a cure-all for society's ills, encouraging us all to explore our relationship to technology. In one of his better-known essays from this period, he warned, "Everywhere we remain unfree and chained to technology, whether we passionately affirm or deny it. But we are delivered over to it in the worst possible way when we regard it as something neutral; for this conception of it . . . makes us utterly blind to the essence of technology."[27] In other words, to the extent that we believe technology to be value-free and without political or symbolic power, we not only misunderstand the very nature of technology, we give it even more sway than it would otherwise have.

In the final analysis, then, it is this symbolism and the effects of technology that give it its power, not the engineering details. To answer the question of why the towers fell, then, requires knowledge not only of the mechanics of materials but also of the dynamics of our interactions with what we create. How can we move on and learn from this? How should we build new buildings, drawing knowledge from the rubble? We can begin by taking a cue from Leslie Robertson, who recognized the complexity of his profession and was pained by the reality.

> It would be good to conclude this journey in a positive mode. We have received almost a thousand letters, e-pistles, and telephone calls in support of our designs. The poignant letters from those who survived the event and from the families of those who both did and did not survive cannot help but bring tears to one's eyes. They have taught me how little I know of my own skills and how fragile are the emotions that lie within me. Yes, I can laugh, I can compose a little story . . . but I cannot escape.
>
> Do those communications help? In some ways they do; in others, they are constant reminders of my own limitations. In essence, the overly laudatory comments only heighten my sense that, if I were as farseeing and talented as the letters would

have me be, the buildings would surely have been even more stalwart, would have stood even longer . . . would have allowed even more people to escape.

Yes, no doubt I could have made the towers braver, more stalwart. Indeed, the power to do so rested almost solely with me. The fine line between needless conservatism and appropriate increases in structural integrity can only be defined after careful thought and consideration of all of the alternatives. But these decisions are made in the heat of battle and in the quiet of one's dreams. Perhaps, if there had been more time for the dreaming. . . .[28]

FOR FURTHER EXPLORATION

1. Look over your calendar or planner for the last week. What items on it required that you use or develop your logical, scientific skills? What items on it required you to use or develop your creative skills? What other skills did you draw on this past week? Which skills did you draw on most often? Which were you most comfortable using? What could you do in the coming week to work on the skills you use less often or are less comfortable with?
2. Identify a building in your neighborhood that you visit regularly (e.g., a classroom building, library, grocery store, shopping mall, fitness center). How do the size, shape, and layout of the building affect how you use it? Does the building's design encourage lingering or efficient pass-through? Are you likely to interact with other people in this building? Are such interactions encouraged by the space? Does it serve its intended purpose well? Can you think of a different building with the same purpose (e.g., another classroom building) that works quite differently? What do you like about the building? What would you change?
3. Choose an iconic structure in a city or town you know well (e.g., city hall, state capitol building, sports arena, campus administration building). What makes the building stand out—size, shape, color, location, style? Are these characteristics functional or decorative? In what ways does the

structure complement or detract from its surroundings? Is it well liked? Why or why not? If you had to replace the structure (following a major fire or earthquake, for example), would you change anything about it or rebuild it as is?

Conclusion

"More Time for the Dreaming": Engineering Curricula for the Twenty-First Century

We humans are born creatures of the earth and air, capable of functioning with our heads in the clouds—so long as our feet remain on the ground.

ERIC DARTON

So what are the characteristics of engineering? It is interconnected, complex, ambiguous, passion-filled, messy, people-oriented, and ultimately, hard. It is hard not so much because it is filled with equations, but because it is filled with equivocations. That is not to suggest engineering is misleading, yet without respect for the complexity of the endeavor, one can be misled into faulty conclusions and false security. Learning the technical material is by far the most manageable part of the process of becoming an engineer. More challenging is to master the nuances and implications and subtleties of what makes for good (that is, efficient, appreciated, sustainable, appropriate, safe) design. The lessons described in this book require reflection and continual refinement. A good and committed student of engineering (which I take to include current practitioners as well as undergraduates) must be willing and able not only to judge a design, but also to adjust to new ideas about design.

In the days and weeks following September 11, 2001, many faculty members struggled with questions of whether and how to address the terrorist attacks with their classes. Should we stick to our syllabi and allow students to make sense of the horrific and heroic events of that day elsewhere? Should we use class time to provide our students a safe place to share their reactions? Should we find a way to use the planned course material as a lens for examining a day that will shape our lives (and perhaps our disciplines) for years to come? Many of us were too stunned to know how to respond that awful Tuesday morning, and for some, sticking to the planned syllabus was comforting: it provided a sense of normalcy and control so desperately needed. Encouraging students to use the classroom as a sort of impromptu counseling center struck some as inappropriate as well as uncomfortable; most of us are untrained to handle such a role.[1]

Yet however ill prepared we may have felt as instructors, our students were even less equipped. My first-year students, for example, had left home just two weeks earlier, and the majority were living on their own for the first time, the newest residents of a campus of over 40,000 students. As their instructors, we were among the few mature adults those students saw on a regular basis. If our role is not merely to instruct students in the finer points of our disciplines but to prepare them to be clear-thinking, well-informed citizens and professionals, we are obliged to deal with world events in the classroom. Granted, this is a position that could be pressed to its absurd extreme, with faculty and students doing little more than discussing the day's news, but I think there is a reasonable middle ground in which we pause to consider engineers' role in its broader context.

It has been twenty years since philosopher Paul Durbin discussed how to help engineering students develop the "wise and prudent foresight" necessary to limit the risks associated with "modern advanced technology."[2] Among the methods proposed by Durbin are better enforcement and teaching of ethics codes, and more integrated humanities and social science training for engineers. Unfortunately, such subjects remain largely relegated to nontechnical courses in the engineering curriculum, freeing the majority of engineering faculty from placing qualitative and ethical questions of risk and safety front and center. How differently might we view the profession if we thought of curricula not in terms of engineering *and* the humanities, but in terms of engineering as *including* the humanities?

Engineers should not be expected to become experts in politics, religion, or international relations, but the terrorist attacks should highlight for us the importance of learning about world events, of being aware of the *effects* of politics and religion and international relations on the practice of engineering. If engineers are to fulfill their "personal obligation to the profession," they must be able to place that obligation in context. What will the code of ethics—particularly the first canon—mean in the twenty-first century? We need not and cannot answer this question definitively, but we can raise it as a topic for consideration.

This book began with "six deceptively simple lessons." First, engineering theory speaks of certitude, but engineering practice is characterized by ambiguity. Second, engineers strive for perfection, and yet they must expect mistakes. Third, engineering expertise is often necessary and yet insufficient to solve the problems of the world. Fourth, engineers value public safety, but they must balance this ideal with other needs. Fifth, engineers succeed through commitment, hard work, and allegiance, but these habits are best when practiced with restraint and moderated by other influences. And finally, engineering curricula tend to reward memorization and conformity, but engineering design requires a balance of creativity and conservatism.

With the complexity and nuance of the preceding chapters in mind, those six lessons or characteristics of engineering lead to three goals for engineering education. First, engineers must learn to expect failure and prepare for success. Second, engineers must learn to be creative in thinking of solutions and conservative in executing them. Third, engineers must develop a deep technical knowledge on which to base designs and a broad social knowledge to situate that design in its context.

Each of these paradoxes is, in fact, reflected in the self-contradictory way we educate engineers. As engineering educators work to resolve these contradictions, engineering students can help themselves by recognizing the disparities in what we ask of them and the ethical dilemmas behind these disparities that make them so difficult to address.

First, educators tell students to challenge themselves by stepping outside their comfort zones, yet set standards (for departmental admissions, job applications, and the like) that demand success. We have yet to identify a reliable, efficient way of engaging in a detailed level of analysis of students' talents. Sometimes the best students are those who have experienced failure and learned from it, rather than those with a perfect 4.0. And yet we tend to rely on grade point averages and other quantifiable assessments of ability, flawed though we know them to be. Students can assist faculty and employers by learning how to assess and sell themselves through measures more complex than the numbers on their transcripts. One sign of a mature learner is what education scholars call "metacognition"—the ability to assess one's own level of knowledge and identify what is needed to improve. It is the rare engineering course that asks students to take time to reflect on what they are learning, how it builds on what they already know, and what more they need to learn, but this sort of reflection is a powerful amplifier of a college education.[3]

Second, observant students will note a tension in their coursework between emphasizing rote mastery of fundamentals (as in problem sets and cookbook labs) and expecting creative thinking (as in problem-based learning, most design courses, and nearly all humanities courses). At least some educators recognize that this tension comes from a dilemma posed by educational theorists, who note that "expertise" is characterized by a mastery of fundamentals *and* the breadth of experience to think accurately and creatively about the problems at hand.[4] In short, to do engineering well requires a balance of structured and creative thinking (in truth, the same is true of all disciplines). As educators explore better ways to weave these two needs together, engineering students can further their own learning by consistently pushing themselves both to delve deeper and to explore more broadly, recognizing that both types of knowledge are crucial to their ultimate success as engineers.

Third, the typical engineering curriculum is packed with technical courses at the "core," sandwiched between design at the first and last year, and sprinkled with liberal studies throughout. Co-ops are for many students the one place where the diverse skills and knowledge of an engineering curriculum come together with any kind of coherence. And yet a curriculum that consisted solely of co-ops would be an apprenticeship, with little of the theoretical base that distinguishes engineers as professionals. We now know that experts differ from novices not only in how much they know, but also—and more importantly—in how they structure what they know. Expertise lies partly in the breadth of knowledge, but also in the connections and parallels and patterns experts identify and use to categorize or organize that knowledge. Although this is a skill that takes time to acquire, it is one that should be striven for throughout the educational process. Once again, educators must provide students (and students must seek) "more time for the dreaming"—for the unstructured, undirected reflection from which the deepest insights emerge.

Exactly what faculty teach students about September 11 is up to individual instructors. What engineers must agree on is that that day will affect future engineering practice and that students must wrestle with the implications of terrorism in order to be prepared to practice in the twenty-first century. Whatever one's views on discussing current events in the classroom, the tragedy of September 11 affords an excellent opportunity for reflection on the very nature of the engineering enterprise.[5]

In the days following the 11th of September 2001, Leslie Robertson lamented what the outcome of that day might have been had he done a bit more. A year later, I found myself wondering what my students might have learned if I, too, had done a bit more, if I had pushed them to see the complexities hidden in the 9/11 documentary we had watched. Each year, the anniversary of 9/11 provides us with an opportunity not just to remember, but also to reflect on the nature of the engineering enterprise. The ultimate lesson that we might glean from such reflection is that context matters—in education as well as in design. This book is the result of my attempts to "do a bit more."

At a minimum, I hope to leave my students with memories of the tragedy that enable them to move past shock and dismay and toward a productive resolve. I want to encourage them to seek an understanding of the events of 9/11 and of engineering design that goes beyond a technical understanding. I also want them to expect their education to provide them opportunities to develop that broader understanding. In the words of Leslie Robertson, "It's a tremendous responsibility, being an engineer."[6] So too is it a tremendous responsibility becoming one.

FOR FURTHER EXPLORATION

Early on, we considered "design under constraint" as a possible definition of engineering. How would you define it now?

Acknowledgments

I am honored to thank the Linda Hall Library in Kansas City, Missouri, whose generous financial and scholarly support facilitated the early stages of this project. My 1995 fellowship at LHL provided me with the resources and the time to complete the first full draft of my dissertation, without which everything that came after would have been impossible.

I am indebted to two professional engineering societies for permission to reprint material previously published in their journals. Portions of this book are based on the following articles, which are used here with permission of the Institute of Electrical and Electronics Engineers: "'A New Era': The Limits of Engineering Expertise in a Post-9/11 World," by Sarah K. A. Pfatteicher, pp. 1–4 in IEEE *International Symposium on Technology and Society (ISTAS) Proceedings*, Las Vegas, NV, 1–2 June 2007 (© 2007 IEEE), and "Learning from Failure: Terrorism and Ethics in Engineering Education," by Sarah K. A. Pfatteicher, *IEEE Technology and Society Magazine* 21 (Summer 2002): 8–12, 21 (© 2002 IEEE).

Portions of Chapter 2 are based on the following articles and are used here with permission of the American Society of Civil Engineers: "'The Hyatt Horror': Failure and Responsibility in American Engineering," by Sarah K. A. Pfatteicher, *Journal of Performance of Constructed Facilities* 14 (May 2000): 62–66 (© 2000 ASCE), and "Walkways: Tragedy and Transformation in Kansas City," by Sarah K. A. Pfatteicher, pp. 47–56 in *Forensic Engineering Conference Proceedings* (2000) (© 2000, ASCE).

I began my study of engineering disasters by delving into the complexities of the Kansas City Hyatt Regency collapse. Eventually I came to appreciate that there are no simple answers to how such tragedies come to pass, much less how to recover from them. Jack Gillum and Dan Duncan know how challenging our early conversations were, and I am deeply touched that they bore with me as I struggled to understand what they have lived.

Apart from my fascination with the subject itself, students are why I do this work. Their energy, curiosity, and potential brighten every September for me when they arrive on campus; and their accomplishments, both personal and scholarly, amaze and inspire me every May. I am blessed that they are too many to name one by one.

Through the years, as students have come and gone, my colleagues at the University of Wisconsin and around the country have contributed in ways large and small, direct and indirect. Ron Numbers, Eric Schatzberg, and Colleen Dunlavy pushed and nurtured me in the early days of this project, and their questions echo in my head to this day—what better compliment can a teacher receive? My fellow graduate students from those days—Ralph Drayton, Libbie Freed, Judy Houck, Craig McConnell, Dave Reid, Michael Robinson, Lisa Saywell, Brewer Stouffer, and all the rest—formed the first academic community I knew, and they continue to set the standard.

Pat Farrell, Jay Martin, and Kathy Sanders each played a key role in developing the Introduction to Engineering Design course whose students inspired this volume; I thank them all for their foresight and for inviting me to join and ultimately lead the class. I have never known a course to attract such a universally talented and committed teaching team—faculty, staff, and student assistants alike—and I am honored to have been a part of it.

Jo Handelsman, Daniel Kleinman, and Bob Ray—all of whom, ironically, were or are in the College of Agricultural and Life Sciences, my new home—have each offered kind words at just the right time that encouraged me more than they can know. Another CALS colleague, Teri Balser, asked me a ridiculous question in late summer 2007, and the resulting project gave me new hope at a difficult time. Joe Herkert at Arizona State University, Bob Ladenson and his colleagues at the Illinois Institute of Technology, and Aarne Vesilind at Bucknell University opened their doors to me and nudged me along, while Clark Miller,

Linda Hogle, and others from the Holtz Center for Science and Technology Studies provided community and friendship closer to home. No matter where she is, Vanessa Northington Gamble continues to amaze and amuse. Mark Mastalski shattered my expectations, and I couldn't be happier about it. I hope they all know that the little things really have mattered.

ASEE LED members routinely provide a forum for testing, discussing, and refining ideas related to engineering history, ethics, communication, philosophy, and education. ASCE and IEEE members have welcomed me into their fold, and their curiosity has been inspirational. What a delight to have colleagues and friends from such far-flung fields!

The women in the group known (more facetiously with every passing year) as the "young Smithies" inspire me ever upward with the variety and creativity of their lives and dreams. In response to my rather desperate phone call from out of the blue back in 1995, Rachel Brown went above and beyond the call of duty of a fellow Smith alumna in welcoming me into Kansas City, her home, and her family. Sophia would be proud of you all.

Bob Brugger, my editor at Johns Hopkins University Press, bore with me for more years than anyone had a right to expect of him. I shall be grateful for far longer than this volume was in coming. Carolyn Moser's excellent copyediting improved the final product in countless ways. If I had known how painless the final stages would be, I might have finished sooner.

I cannot begin to repay Jeff Russell and Mike Corradini, without whom I would not have found a long-term home at UW–Madison. Instead, I try every day to live up to their example and their integrity. Alice Pawley, Jen Kushner, and Maggie Tongue provide sustenance to body and spirit no matter what state they are in, but I do miss those days together on the fourth floor. Judy Houck and Lisa Saywell have made the scholarly life—and life in general—less isolating and more entertaining. How wonderful that they found their way home to Madison. I am lucky indeed to count such folks as friends and colleagues.

Peter Bosscher and Brian "Scoop" O'Connell each epitomized what it means to be a good colleague and a better friend, and I miss them both dearly.

Philip Pfatteicher and Lois Sharpless Pfatteicher introduced me to the notion that faith and thinking are not contradictory, and encouraged me to sort the rest out for myself. I hope they each see their influence in these pages.

And most of all, an enormous thank-you to Bob Conlin and Ian Pfatteicher Conlin, for everything. I love you. Sorry for saying so in public.

Notes

Introduction: Why?

1. "Continuing Lessons of 9/11," *New York Times,* 20 May 2004, p. A26.

2. *Faith and Doubt at Ground Zero*, videocassette, dir. Helen Whitney (Frontline, PBS, 2002); Yosri Fouda and Nick Fielding, *Masterminds of Terror: The Truth behind the Most Devastating Attack the World Has Ever Seen* (New York: Arcade, 2003); Therese McAllister, ed., *World Trade Center Building Performance Study: Data Collection, Preliminary Observations, and Recommendations,* FEMA Report 403 (Washington, DC: FEMA, Federal Insurance and Mitigation Administration, May 2002) (hereafter, *Building Performance Study*). Another example of the search for a spiritual explanation is Amy Bartlett, *Be Still, America . . . I Am God: From Out of the Rubble, Stories of Hope* (Camp Hill, PA: Christian Publications, 2001). Other works exploring the political and social background of the attacks are National Commission on Terrorist Attacks, *The 9/11 Commission Report: Final Report of the National Commission on Terrorist Attacks upon the United States* (New York: Norton, 2004) (hereafter *9/11 Commission Report*); Simon Reeve, *The New Jackals: Ramzi Yousef, Osama bin Laden and the Future of Terrorism* (Boston: Northeastern University Press, 1999); Gilbert Achcar, *The Clash of Barbarisms: September 11 and the Making of the New World Disorder*, trans. Peter Drucker (New York: New York University Press, 2002); Richard Crockatt, *America Embattled: 9/11, Anti-Americanism and the Global Order* (New York: Routledge, 2003); and Thomas L. Friedman, *Longitudes and Attitudes: The World in the Age of Terrorism*, updated and expanded ed. (New York: Anchor, 2003).

3. For a notable reference to the comfort offered by technical analyses, see Leslie E. Robertson, "Reflections on the World Trade Center," *The Bridge* (National Academy of Engineering) 32 (Spring 2002): 5–10.

4. Body of Knowledge Committee of the Committee on Academic Prerequisites for Professional Practice, *Civil Engineering Body of Knowledge for the 21st Century*, 2nd ed. (Reston, VA: American Society of Civil Engineers, 2008); National Academy of Engineering, *Engineer of 2020: Visions of Engineering in the New Century* (Washington, DC: National Academies Press, 2004); *Report of the Engineering Licensure Qualifications Task Force* (Clemson, SC: National Council of Examiners for Engineering and Surveying, 2003); Accreditation Board for Engineering and Technology, *Sustaining the Change* (Baltimore, MD: Accreditation Board for Engineering and Technology, 2004); Wallace Fowler, "Tomorrow's Engineering Education; President's Message," *PRISM* (ASEE) 10 (January 2001): 38; and "President Annual Message: Engineering Education in the Next Decade; The Challenge of Change" in *ASEE Annual Report for 2000*, www.asee.org/about/annualreport2000.cfm.

5. Eric Klinenberg, *Heat Wave: A Social Autopsy of Disaster in Chicago* (Chicago, IL: University of Chicago Press, 2002), p. 23; Steven Biel, ed., *American Disasters* (New York: NYU Press, 2001), p. 5.

6. Eric Darton, "The Janus Face of Architectural Terrorism: Minoru Yamasaki, Mohammad Atta, and Our World Trade Center," in *After the World Trade Center: Rethinking New York City*, ed. Michael Sorkin and Sharon Zukin, p. 95 (New York: Routledge, 2002). Whether the dominance of INTJs is a necessary characteristic of engineering or whether the current structure of engineering tends to encourage those with INTJ traits is a worthwhile question. For our purposes here, it is enough to understand that the current makeup of the field is overwhelmingly INTJs (as it is overwhelmingly white and Asian males).

7. Langdon Winner, "Do Artifacts Have Politics?" in *The Whale and the Reactor: A Search for Limits in an Age of High Technology (*Chicago: University of Chicago Press, 1986); Michael Martin and Roland Schinzinger, *Ethics in Engineering*, 3rd ed. (New York: McGraw Hill, 1996).

8. Charles Perrow, *Normal Accidents: Living with High-Risk Technologies* (Princeton, NJ: Princeton University Press, 1999); Eric Darton, *Divided We Stand: A Biography of New York's World Trade Center* (New York: Basic Books, 1999). See also Michael Davis, *Thinking Like an Engi-*

neer: Studies in the Ethics of a Profession (New York: Oxford University Press, 1998).

One: "A Very Imperfect Process"

Epigraph: Leslie E. Robertson, "Reflections on the World Trade Center," *The Bridge* (National Academy of Engineering) 32 (Spring 2002): 7.

1. "Testimony of Dr. W. Gene Corley on Behalf of the American Society of Civil Engineers before the Subcommittee on Environment, Technology and Standards & Subcommittee on Research Committee on Science U.S. House of Representatives March 6, 2002," p. 3, accessed at www.asce.org/pdf/3-6-02wtc_testimony.pdf (hereafter cited as Corley, "Testimony").

2. Therese McAllister, ed., *World Trade Center Building Performance Study: Data Collection, Preliminary Observations, and Recommendations,* FEMA Report 403 (Washington, DC: FEMA, Federal Insurance and Mitigation Administration, May 2002); hereafter cited as *Building Performance Study.* Both FEMA and the ASCE host websites to complement the report, which is available for free in electronic and hard copy. The report is available online at www.fema.gov/library/wtcstudy.shtm. Gene Corley (lead investigator) has also written several synopses of the report and its findings; see, for example, W. Gene Corley, "What We Learned: Building Performance Study of the WTC Collapse," *Structural Engineer Magazine* 3 (August 2002): 26–31. The ASCE published a parallel study of the Pentagon, released early in 2003 and available via the ASCE September 11 website: www.asce.org/responds. The design of the Pentagon has also been credited with saving lives; see, for example, "Experts Say Pentagon's Design Saved Lives," *USA Today*, 24 January 2003, p. A5.

3. Corley, "Testimony," p. 6, estimates that the team's resources equaled about $1 million and that roughly 40 times that amount would have been necessary for a "comprehensive study." See also James Glanz and Eric Lipton, *City in the Sky: The Rise and Fall of the World Trade Center* (New York: Times Books, 2003), pp. 329–33, as well as *Building Performance Study*, pp. 8–13; Angus Kress Gillespie, *Twin Towers: The Life of New York City's World Trade Center*, rev. ed. (New York: New American Library, 2002), p. 248; and William Langewiesche, *American Ground: Unbuilding the World Trade Center* (New York: North Point Press, 2002), p. 51.

4. One engineer (reportedly a member of the team in the early days)

happened to glance out his hotel window just in time to see a load of steel resting on a flatbed truck, headed for the salvage yard. Though he held no official credentials, he simply hurried out and took photos and notes on what he saw (Glanz and Lipton, *City in the Sky*, p. 331, citing Kenneth Chang, "Scarred Steel Holds Clues, and Remedies," *New York Times*, 2 October 2001, p. F1). For more on this engineer, Abolhassan Astaneh-Asl, see Langewiesche, *American Ground*, pp. 50–51. See also Usha Lee McFarling, "Finding Hope in the Ruins" *Los Angeles Times*, 15 November 2001, p. A1.

5. Langewiesche, *American Ground*, p. 96.

6. Thomas W. Eagar and Christopher Musso, "Why Did the World Trade Center Collapse? Science, Engineering, and Speculation," *JOM: The Journal of the Minerals, Metals and Materials Society* 53 (December 2001): 8–11, available online at www.tms.org/pubs/journals/JOM.

7. Samuel C. Florman, *The Existential Pleasures of Engineering*, 2nd ed. (New York: St. Martin's Griffin, 1994), p. 177. See, too, Louis L. Bucciarelli, *Designing Engineers* (Cambridge, MA: MIT Press, 1994), p. 48, on simplifying and classifying the natural world as providing scientists with "a sense of control."

8. From a review of *NOVA*'s *Why the Towers Fell* on Amazon.com at www.amazon.ca/gp/cdp/member-reviews/A1E9AE0X4B6UYA, posted 29 April 2003 by Sophie M. "Nessinette" of San Francisco.

9. Robertson, "Reflections on the World Trade Center," p. 7.

10. "Continuing Lessons of 9/11," *New York Times*, 20 May 2004, p. A26. The national commission referred to was the federal 9/11 Commission.

11. The *New York Times* cited insurance claims by WTC leaseholder Larry Silverstein and lawsuits by victims and their families as drivers for some investigations (most notably the one commissioned by Silverstein himself) and the awkwardness that such inquiries causes (James Glanz and Eric Lipton, "Expert Report Disputes US on Trade Center Collapse," *New York Times*, 22 October 2002, p. B1). See, too, an article in *Engineering News-Record* reporting that members of the Structural Engineers Association of New York "are concerned about the possible impact [of misinterpretations of engineering studies] on a lawsuit filed against the port authority" (Nadine M. Post, "Engineers Set the Record Straight on Trade Center Study Results," 4 November 2002, *Engineering News-Record*, online at http://enr.construction.com/news/buildings/archives/021104a.asp; see also

Post, "Study Absolves Twin Tower Trusses, Fireproofing," *Engineering News-Record*, 4 November 2002, p. 12). And see James Glanz, "Comparing Two Sets of Twin Towers; Malaysian Buildings Offered as Model," *New York Times*, 23 October 2002, p. B1, who wrote, "Many technical experts say there are good reasons they have dealt so gingerly with such a volatile, if perhaps valuable, question," including concern for feelings of victims' families, desire not to give information to terrorists, and the "sheer complexity of actually producing an answer."

12. See, for instance, Caroline Whitbeck's *Ethics in Engineering Practice and Research* (Cambridge: Cambridge University Press, 1998), and Lisa Newton's D.I.S.O.R.D.E.R. schema, available under "Manuals" at www.rit.edu/cla/ethics. Engineer and author Samuel Florman has famously written of his conviction that engineering skills are valuable more broadly; see, for example, his chapter "Of Dullards and Demigods" in *Existential Pleasures*.

13. A more detailed introduction to what forensic engineers do is provided in Geetha Rao, "Anatomy of an Accident: How Forensic Engineers Determine What Went Wrong," *Risk Management* 42 (October 1995): 63–65.

14. *Building Performance Study*, p. 1.

15. Corley, "Testimony," p. 6; Langewiesche, *American Ground*, pp. 11–12, 51, 55, 192; Glanz and Lipton, *City in the Sky*, pp. 329–33. Also, *Building Performance Study*, pp. 8–13, and Gillespie, *Twin Towers*, p. 248. Langewiesche notes the towers were roughly 90% air and 10% structure, which would equal the eleven stories of compacted debris; the structure itself weighed roughly 600,000 tons, and the structure and contents amounted to 1.5 million tons of rubble, "tied together like steel wool."

16. Langewiesche, *American Ground*, pp. 94–95.

17. Ibid., pp. 210, 113, 9. For more on the management structure on the site, see also pp. 68–69, 88–89, 90–96, 108–9, 113, 118, 128, 171, 178, 190.

18. Ibid., pp. 18, 119, 128. For more information on the activities of engineers in the early days, see Cathy Murphy, "Called to Action: How Structural Engineers Respond to Disasters," *Structural Engineer Magazine* 2 (January 2002): 36–41, and "Group Developing Emergency Response Plan for Structural Engineers" *Structural Engineer Magazine* 2 (January 2002): 13. Note, too, that the National Science Foundation funded a variety of researchers' explorations of the causes and effects of the collapse; see

"Grants Awarded to Study Attacks" and "Georgia Tech Professor Invited to Study World Trade Center Damage," both in *Structural Engineer Magazine* 2 (January 2002): 10. For one investigator's description of the early days on-site, see Abolhassan Astaneh-Asl, "Testimony Before the Committee on Science of the U.S. House of Representatives," March 6, 2002 Hearing, *Learning from 9/11: Understanding the Collapse of the World Trade Center*, at http://thewebfairy.com/nerdcities/WTC/astaneh-wtc.htm.

19. Corley, "Testimony," p. 6.

20. Ibid., p. 3. See also Langewiesche, *American Ground*, p. 12, and p. 107 on the "grim sort of hope" that drove workers on the pile; Glanz and Lipton, *City in the Sky*, p. 330. The *Building Performance Study*, chap. 1, pp. 8–14, reports that the initial response of the engineering community was to (1) ensure rescue and recovery could proceed safely and (2) assess whether and when damaged surrounding buildings could be reoccupied. Also see *Building Performance Study*, app. F, pp. 1–2, on recommendations for future emergency response (to avoid problems encountered in this one).

21. Langewiesche, *American Ground*, pp. 110, 193–96, 203–5, for information on salvage and recycling activity at Fresh Kills. On engineers' difficulty in getting access to the steel, see James Glanz and Kenneth Chang, "Engineers Seek to Test Steel before It Is Melted for Reuse," *New York Times*, 9 September 2001, p. B9. Corley, "Testimony," p. 5, provides information on the earliest days of the study team's work on site; on p. 6 Corley denies that that team's work was "hampered because debris was removed from the site and has subsequently been processed for recycling." Rather, Corley testified, the team "had full access to the scrap yards and to the site and has been able to obtain numerous samples. At this point there is no indication that having access to each piece of steel from the World Trade Center would make a significant difference to understanding the performance of the structures."

22. Dates on site from Corley, "Testimony," p. 5, which also says: "Unlike other structural collapses, there is an unprecedented volume of photographic and video evidence available for the team to review, including more than 120 hours of network and private video footage. Individual team members have viewed every foot of this videotape and provided information on the available data to the team at large." Also *Building Performance Study*, p. 2 and app. E; Glanz and Lipton, *City in the Sky*, p. 269.

23. Jim Dwyer and Kevin Flynn, *102 Minutes: The Untold Story of*

the Fight to Survive inside the Twin Towers (New York: Times Books, 2005), p. 212; see also pp. 242–43 for an additional description of the moments of the collapse.

24. Both excerpts are from Langewiesche, *American Ground*, p. 53.

25. Dwyer and Flynn, *102 Minutes*, p. xxi; Claire Coleman, "Gene Corley: Senior Structural Engineer Who Led the Official Investigation into Collapse of the Twin Towers," *The Guardian*, 11 September 2002, www.guardian.co.uk/world/2002/sep/11/september112002.september1196.

26. The Controlled Demolition, Inc., website has information on the company's participation in studying the Murrah Federal Building in Oklahoma City, at www.controlled-demolition.com/default.asp?reqMode=1&reqLocId=7&reqItemId=20030317124730.

27. Langewiesche, *American Ground*, pp. 51, 54–55.

28. *Building Performance Study*, chap. 2, esp. pp. 27 and 35; Langewiesche, *American Ground*, pp. 58–61.

29. Coleman, "Gene Corley."

30. "Cleanup Effort Ends at World Trade Center as Collapse Report is Released," *Structural Engineer Magazine* 4 (June 2002): 17.

31. *Building Performance Study*, pp. 2–38. See also Corley, "What We Learned," and Jennifer Goupil, "More to Learn: Significant Findings from the WTC Investigation," *Structural Engineer Magazine* 4 (August 2002): 31.

32. Langewiesche, *American Ground*, pp. 51–53.

33. Langewiesche, *American Ground*, p. 53.

34. James Glanz "Why Trade Center Towers Stood, Then Fell," *New York Times*, 11 November 2001, p. B1; Langewiesche, *American Ground*, p. 56; Nadine M. Post, "After Disaster: The Three R's of Threat Design—Resist, Respond, Recover—Are Catapulted to the Front Lines," *Engineering News-Record*, 31 December 2001, p. 8; Council on Tall Buildings and Urban Habitat, "Task Force on Tall Buildings: 'The Future,'" October 15, 2001, pp. 17 and 46, accessed at www.ctbuh.org.

35. Dwyer and Flynn, *102 Minutes*, p. xxiii. Andrew Hazucha, "Chicago on the Brink: Media Trauma and the 1977 L-Train Crash," in *American Disasters*, ed. Steven Biel (New York: New York University Press, 2001), describes the city's evolving reactions to the train crash that killed 11 and injured 266. He postulates that Chicagoans were able to come to terms with that tragedy only by accepting it "not as a harbinger of the apocalypse, but rather as an inevitable event in a world made dangerous by technol-

ogy." An inevitable event (even if caused by human error) is, according to Hazucha (and Freud, whose research he cites), more manageable by the human psyche because it produces only ongoing anxiety rather than the outright, crippling fear that results from the truly unexpected. By analogy, then, one might expect the public to be better able to cope with the idea that the WTC collapse had resulted from flaws in design or evacuation (i.e., human shortcomings with which we can all identify) rather than from evil incarnate in the form of terrorists intent on doing harm (a mindset many of us have difficulty conceiving).

36. Dwyer and Flynn, *102 Minutes*, pp. 252, 254. Also on fireproofing, see "Investigators Question Trade Center Towers' Construction Specifications," *Structural Engineer* 4 (2003): 10–11, and James Glanz, "Towers Untested for Major Fire, Inquiry Suggests," *New York Times,* 8 May 2003, p. A1. On building motion studies, see Glanz and Lipton, *City in the Sky*, pp. 140–44, 323–24; and Etienne Benson, "The Unexpected Benefits of Basic Science," *Monitor on Psychology* 34 (January 2003): 36–37. Detailed studies of the evacuation of the Twin Towers were conducted following the 1993 bombing; see, e.g., Rita F. Fahy, "The Study of Occupant Behavior during the World Trade Center Evacuation: Preliminary Report of Results," pp. 196–202, in *Proceedings of the International Conference on Fire Research and Engineering*, September 10–15, 1995, Orlando, FL.

37. *Why the Towers Fell*, videocassette, prod. and dir. Garfield Kennedy and Larry Klein (*NOVA*, WGBH, 2002); quotations from transcript, pp. 1 and 22. The *New York Times* reported on 23 October 2002 that Thornton "is generally careful to say he believes Manhattan's twin towers were built as well as they could have been *in their day*" (Glanz, "Comparing Two Sets of Twin Towers"; emphasis added). The *Times* reported in this same article that Thornton had used the World Trade Center primarily to highlight the features of his own building, a fit of insensitive and regrettable grandstanding.

38. Glanz, "Comparing Two Sets of Twin Towers."

39. Dwyer and Flynn, *102 Minutes*, pp. xxii.

40. On the difficulty of raising questions about whether the terrorists were solely responsible, see, for example, coverage of and reaction to the description by University of Colorado professor Ward Churchill of some of the 9/11 victims as "little Eichmanns" who were complicit in their fate, reported in Michelle York, "Professor Quits a Post over a 9/11 Remark," *New York Times,* 1 February 2005, p. B8, and Dahlia Lithwick, "Stupidity

as a Firing Offense," *Slate Magazine*, 10 February 2005, www.slate.com/id/2113358; Jon Magnusson, "Speak Up!" *Civil Engineering* 74 (March 2004): 8 (emphasis in original).

41. Christopher M. Foley, "Why They Fell," *PRISM* (ASEE)11 (December 2001): 6–7.

42. *World Trade Center: Anatomy of the Collapse*, videocassette, prod. David Darlow and dir. Ben Bowie (Artisan Home Entertainment, 2002), transcript, p. 12.

43. Ibid., p. 1.

44. Ibid., pp. 2, 8, 10, 12, 7, 2; *Why the Towers Fell,* transcript, p. 23. Larry Silverstein, landlord of the World Trade Center, sponsored a separate study, which also concluded that the engineers were not guilty but disputed FEMA's findings. See Glanz, "Comparing Two Sets of Twin Towers," for a description of the Silverstein report and Post, "Engineers Set the Record Straight," for clarification of the differences between the reports.

45. "Continuing Lessons of 9/11," *New York Times*, p. 26; John Seabrook, "The Tower Builder," *New Yorker*, 19 November 2001, p. 73.

Two: "Finding Hope in the Ruins"

Epigraph: Quoted in Usha Lee McFarling, "Finding Hope in the Ruins," *Los Angeles Times*, 15 November 2001, p. A1.

1. McFarling, p. A1. For a lively review of some classic literature on failures, see Henry Petroski, "The Success of Failure," *Technology and Culture* 42 (April 2001): 321–28.

2. "The Falling of the Dixon Bridge," *Scientific American* 28 (24 May 1873): 321.

3. Ibid. As Henry Petroski has pointed out, "Exactly how many railway bridges failed in the nineteenth century is not so important as the fact that failures did occur." Petroski concludes, quite rightly, that "the *perceived* risk of failure . . . was certainly high." See Petroski, "On Nineteenth-Century Perceptions of Iron Bridge Failures," *Technology and Culture* 24 (1983): pp. 655 and 658; emphasis in original. Historian David McCullough estimates that "something like forty bridges a year fell in the 1870s—or about one out of every four built. In the 1880s some two hundred more fell. Highway bridge failures were the most common, but the railroad bridge failures received the greatest publicity and cost the most lives." David McCullough, *The Great Bridge* (New York: Simon

and Schuster, 1972), p. 390, citing Joseph Gies, *Bridges and Men* (Garden City, NY: Doubleday, 1963), pp. 125–30. The rate of "not less than forty bridges . . . every year" is also given by George L. Vose, *Bridge Disasters in America: The Cause and the Remedy* (Boston: Lee and Shepard, 1887), p. 3; Vose also notes that disasters were "altogether too common in the United States" (p. 11). Also on the rate of bridge failures in the nineteenth century, see Carl W. Condit, *American Building Art: The Nineteenth Century* (New York: Oxford University Press, 1960), who cites J. A. L. Waddell, *Bridge Engineering* (New York: Wiley, 1916), 2:1539–40. For a debate and brief overview of the past attempts and problems associated with calculating the rate, see an exchange in *Technology and Culture* that began with Sara Ruth Watson's review of Howard Miller's *The Eads Bridge* (21 [1980]: 97–100) and continued with several letters to the editor (22 [1981]: 846–50). The exchange culminated with Petroski's 1983 article "On Nineteenth-Century Perceptions of Iron Bridge Failures," cited above.

4. "The Falling of the Dixon Bridge," p. 321.

5. James B. Eads et al., "On the Means of Averting Bridge Accidents," *ASCE Transactions* 4 (1875): 122. This study is abstracted in Vose, *Bridge Disasters*, pp. 22–27.

6. Petroski, "On Nineteenth-Century Perceptions," p. 656.

7. Eads, "Means of Averting Bridge Accidents," p. 128n.

8. Ibid., pp. 132, 129.

9. Gies, *Bridges and Men*, pp. 125–32, esp. 126–27. See Gabriel Leverich, "Notes and Comments on the Failure of the Bridge on the Lake Shore and Michigan Southern Railway over Ashtabula Creek, December 30, 1878," a bound volume of newspaper and professional journal clippings with handwritten notes in the Engineering Societies Library Collection at the Linda Hall Library in Kansas City, Missouri. Leverich prepared the clippings volume as support for his motion at an ASCE meeting that the society should discuss the causes of failures such as the Ashtabula Creek accident. Vose, *Bridge Disasters*, notes that the railway company paid "over half a million dollars in damages" (p. 70).

10. Leverich, "Notes and Comments," including, for example, *Engineering News* clipping (labeled "Engineers Statement-sinking") from 6 January 1877, p. 8, on the early theory; "Towlinson Opinion," clipping from the *New York Sun*, 12 January 1877; "Towlinson, Beebout's Opinions, Ohio Legislature," *New York Tribune*, 17 January 1877; and "Beebout's Opinion," *New York Times*, 17 January 1877.

11. Quotations from Eric DeLony, "The Golden Age of the Iron Bridge," *Invention and Technology* 10 (Fall 1994): 11. On the same page, DeLony reports that the chief of the Historic American Engineering Record has said that "the Howe truss may be the closest that wooden-bridge design ever came to perfection. For simplicity of construction, rapidity of erection, and ease of replacing parts, it stands without rival."

12. Charles MacDonald, "The Failure of the Ashtabula Bridge," *ASCE Transactions* 6 (1876–77): 74–87. Historian Sara Ruth Watson notes bluntly that "Amasa Stone operated in Cleveland and caused the notorious Ashtabula bridge failure in 1876." Watson asks, "How many of the railroad truss-bridge failures can be laid at the door of Amasa Stone and his interests?" (Watson, review of *The Eads Bridge*, p. 99). The clippings in Leverich, "Notes and Comments" contain assorted disparaging comments about the coroner's inquiry for determining the design (as well as the inspection) faulty.

13. MacDonald, "Failure of the Ashtabula Bridge," pp. 83, 86, 87; "On the Failure of the Ashtabula Bridge (Discussions on Subjects Presented at the 9th Annual Convention)," *ASCE Transactions* 6 (1876–77): 201; Gies, *Bridges and Men*, pp. 125–32, esp. 126–28; McCullough, *The Great Bridge*, p. 390.

14. *Adams Bill*, HR 4538, February 1877; "On the Failure of the Ashtabula Bridge (Discussions)," pp. 202, 207, 210n.

15. "On the Failure of the Ashtabula Bridge (Discussions)," p. 222. For additional coverage of this debate, see "Minutes of Meetings," *ASCE Proceedings* 3 (Sept. 1877): 86–87 (bound with *ASCE Transactions* vol. 6); "Minutes of Meetings," *ASCE Proceedings* 1 (1875): 260ff.; Vose, *Bridge Disasters*, pp. 79–80.

16. Gies, *Bridges and Men*, p. 212. See also Vose, *Bridge Disasters*, pp. 12–16. On a New York City hotel failure, see H. deB. Parsons, "The Collapse of a Building during Construction," and discussion, *ASCE Transactions* 53 (1904): 1–35, which includes a comment that failures result from incompetent engineers, the suggestion that a more severe penalty for incompetence would increase safety, a reference to the Ashtabula failure (which, like the Hotel Darlington collapse described by Parsons, resulted from cast-iron column failure), and a mention of the comfort to be found in the fact that "no member of our profession had anything to do with the designing and construction of this structural monstrosity" (pp. 24–25). On the ASCE's reaction to failures, see William H. Wisely, "In the Spirit of

Public Service," in his *American Civil Engineer, 1852–2002: The History, Traditions, and Development of the American Society of Civil Engineers* (Reston, VA: ASCE, 2002), pp. 245–95.

17. Quotations from Watson, review of *The Eads Bridge,* p. 98. Vose made a similar point about the science of engineering and stated that the safety of a bridge "is not at all a matter of opinion, but a matter of fact and arithmetic" (*Bridge Disasters*, pp. 18, 40–41).

18. George H. Benzenberg, "The Engineer as a Professional Man," ASCE Presidential Address, *ASCE Transactions* 58 (1907): 523.

19. Many authors have written about the Quebec Bridge. One of the more notable works is John Tarkov, "A Disaster in the Making," *Technology and Culture* 1 (Spring 1986): 10–17. A briefer account is given by Gies, *Bridges and Men*, pp. 222–27. See Paul G. Sibly and A. C. Walker "Structural Accidents and Their Causes," *Proceedings of the Institution of Civil Engineers* (London) 62 (May 1977): 191–208; Henry Petroski, "Predicting Disaster," *American Scientist* 81 (March–April 1993): 110–13 (commenting and extrapolating on Sibly and Walker's argument); and Petroski, "History and Failure," *American Scientist* 80 (November–December 1992): 523–52 (suggesting a broader context for the work of Sibly and Walker). See also Cynthia Pearson and Nobert Delatte, "Collapse of the Quebec Bridge," *ASCE Journal of Performance of Constructed Facilities*, 20 (February 2006): 84–91. On a related note, see Sibly's contributions to "The Relevance of History: Open Discussion," *Structural Engineer* 53 (September 1975): 387–98, responding to Alfred Pugsley et al., "The Relevance of History," *Structural Engineer* 52 (December 1974): 441–45.

20. "Bridge Falls, Drowning 80," *New York Times,* 30 August 1907, p. 1 (this early report of the death toll proved to be slightly overstated); Tarkov, "Disaster in the Making," pp. 10–11.

21. See, for example, "Quebec Bridge (Progress Notes)," *Engineering News* 58, pt. 1 (1907): 53; "Quebec Bridge (Progress Notes)," *Engineering News* 57, pt. 1 (1906): 705; "Cantilever Bridge of 1800-Foot Span across the St. Lawrence River," *Engineering News* 56, pt. 2 (1905): 272; "The Quebec Bridge," *Engineering* 84 (13 September 1907): 351–55. The editor added a note to this last article, saying: "This article was, of course, written prior to the occurrence of the grave disaster with which we had occasion to deal in our last issue. As a complete account of a most important work, however, it has lost none of its interest, and a study of the information

which it contains will add much to the understanding of the inquiry now being made into the causes of the failure."

22. "Memoir of Theodore Cooper," *ASCE Transactions* 84 (1921): 828–30; quotation on p. 830. See also "Theodore Cooper Dies in Eighty-First Year," *Engineering News-Record* 83 (28 August 1919): 443. An obituary of Cooper also appeared in the *New York Times*, 25 August 1919.

23. Quotations from Theodore Cooper to E. A. Hoare, 31 July 1903, in Henry Holgate et al., *Report of the Canada Royal Commission on the Quebec Bridge Inquiry*, Sessional Paper 154 (Ottawa: S. E. Dawson, 1908), 1:43 [hereafter cited as *Quebec Bridge* (1908)]; and Tarkov, "Disaster in the Making," pp. 15–16. Also see "A History of the Development of the Specifications and a Discussion of the Evidence Relating to It," app. 6 in *Quebec Bridge* (1908), 1:46; and "Testimony of Theodore Cooper to the Commission," *Quebec Bridge* (1908), 2: 345, 354. In his testimony to the commission, Cooper stated, "As I now see it, I was acting not only as the consulting engineer but as the chief engineer on the Quebec bridge" (app. 6, *Quebec Bridge* [1908], 1:48; see also p. 409).

24. *Quebec Bridge* (1908), 2:345, 353, 357. Cooper repeated his remarks about offering to resign in his testimony to the commission (*Quebec Bridge* [1908], 2:408).

25. "Had Warned Men on Bridge: Theodore Cooper, Consulting Engineer," *New York Times*, 31 August 1907, p. 1.

26. *Quebec Bridge* (1908), 2:357, 413; "Bridge Warning Was Just Too Late," *New York Times*, 1 September 1907, p. 3; telegram, exhibit 79u, in *Quebec Bridge* (1908), 2:539. For further information on the Phoenix Bridge Company and its troubling safety record, see "Phoenix Bridge Company," an essay by the Historical Society of the Phoenixville Area, posted at http://hspa-pa.org/phoenix_bridge.html.

27. *Quebec Bridge* (1908), 1:5; St. Lawrence Bridge Company, *The Quebec Bridge; Carrying the Transcontinental Line of the Canadian Government Railways over the St. Lawrence River near the City of Quebec, Canada* (Quebec, 1918), pp. 2, 21 [hereafter cited as *Quebec Bridge* (1918)]; "Charges Against Cooper," *New York Times*, 21 November 1907, p. 4. A copy of the *Quebec Bridge* (1918) report is in the Linda Hall Library, Kansas City, Missouri. The Canadian Commission's records are held at the National Archives of Canada (NAC), RG 33/6, under the heading "Royal Commission on the Quebec Bridge Inquiry." My thanks to James Whalen, archivist responsible for the textual records of the com-

mission at the NAC, who was most helpful in identifying and locating these documents. See also the inventory to NAC RG 33, titled "Records of Federal Royal Commissions."

28. Quebec Bridge (1908), 1:7, 9; "Findings of the Canadian Commission on the Quebec Bridge," *Engineering News* 59 (12 March 1908): 288–89; "The Report of the Royal Commission of Inquiry on the Collapse of the Quebec Bridge," *Engineering News* 59 (19 March 1908): 307–15; "Charges against Cooper," p. 4; Tarkov, "Disaster in the Making," p. 17; "Memoir of Theodore Cooper," p. 830.

29. Information on the 1916 collapse of the Quebec Bridge can be found in the following: "Giant Bridge Span Collapses; 11 Dead," *New York Times*, 12 September 1916, p. 1; "Quebec Bridge Disaster," *Scientific American*, 116 (23 September 1916): 282–83; "Why Quebec's Bridge-Span Fell," *Literary Digest* 53 (30 September 1916): 827–28; "Further Light on the Quebec Bridge Disaster," *Scientific American* 115 (30 September 1916): 298–99; Norman Carlisle, "Canada's Bridge to Doom: Quebec Bridge," *Coronet* 50 (August 1961): 84–88; "Quebec Bridge: Major Disaster, Major Lesson," *Civil Engineering* 46 (August 1976): 65; Rolt Hammond, *Engineering Structural Failures: The Causes and Results of Failure in Modern Structures of Various Types* (London: Odhams Press, 1956), pp. 110–15; and Gies, *Bridges and Men*, pp. 234–35, 238. The Canadian government commissioned a supplemental report by C. C. Schneider in their 1908 findings to determine the usefulness of the original plans for rebuilding the bridge; the report appeared in vol. 3 of *Quebec Bridge* (1908). *Quebec Bridge* (1918), p. 17.

30. "Why Quebec's Bridge-Span Fell," p. 828, quoting an earlier article in the *Engineering Record*.

31. "Tacoma Narrows Bridge Wrecked by Wind," *Engineering News-Record* 125 (14 November 1940): 647; "Dynamic Wind Destruction," *Engineering News-Record* 125 (21 November 1940): 672–73. Photos of the failure appear in "Details of Damage to Tacoma Narrows Bridge," *Engineering News-Record* 125 (28 November 1940): 720; the famous footage of the failure is announced in "Film Showing Collapse of Tacoma Span Available," *Engineering News-Record* 125 (5 December 1940): 733. The footage itself is readily available online. The record of the effort to understand the collapse of Tacoma Narrows is found in the following: "Board Named to Study Tacoma Bridge Collapse," *Engineering News-Record* 125 (28 November 1940): 725; "Tacoma Bridge Boards Meet in New York,"

Engineering News-Record 126 (23 January 1941): 143; "More Engineers Enter Tacoma Bridge Studies," *Engineering News-Record* 126 (6 March 1941): 372; "Why the Tacoma Narrows Bridge Failed," *Engineering News-Record* 126 (8 May 1941): 743–47; F. B. Farquharson, "Lessons in Bridge Design Taught by Aerodynamic Studies," *Civil Engineering* (ASCE) 16 (August 1946): 344–45. See also Robert R. Goller, "The Legacy of 'Galloping Gertie' 25 Years After," *Civil Engineering* 35 (October 1965): 50–53. A number of technical articles grew out of studies of the Tacoma Narrows Bridge; see, for example, several in the ASCE *Transactions* of 1952, 1953, and 1955.

32. Goller, "Legacy of 'Galloping Gertie'," pp. 51, 53.

33. For a more thorough history of the Hyatt disaster, see *Journal of Performance of Constructed Facilities* (ASCE) 14 (May 2000), which prints a series of retrospective articles on the collapse. For information on two partygoers in particular, see "Talking it Out," *Kansas City Times*, 25 July 1981, p. A4. Four other families were chronicled in the *Kansas City Star*, Special Section, 19 July 1981.

34. General descriptions of the collapse were provided in hundreds of articles in the *Kansas City Star* and *Kansas City Times* in the days and weeks following the collapse. The *New York Times* and other newspapers across the country and around the world also carried the story, but in more abbreviated form. See also Thomas L. Jackson and Ralph L. Kaskell, Jr., "The KC Hyatt Regency Disaster: What Went Wrong?" *Defense Counsel Journal* 56 (October 1989): 415–25 (p. 415 for tonnage). The National Bureau of Standards investigative report gave extensive detail on the events of 17 July 1981, characterizing the collapse as the worst structural disaster in the United States up to that point. See National Bureau of Standards, *Investigation of the Kansas City Hyatt Regency Walkways Collapse*, NBS Building Science Series 143 (Washington, DC: U.S. Department of Commerce, 1982), hereafter cited as *NBS Report*. The *Kansas City Star* used the phrase "The Hyatt Horror" as its headline on 19 July 1981 to open its special section devoted to the disaster. The phrase frequently reappeared in both Kansas City newspapers.

35. For Duncan and Gillum's reaction, see "Hyatt Engineers Found 'Guilty' of Negligence," *Engineering News-Record* 214 (21 November 1985): 10–12 (quotation on p. 10). Jack Gillum stated that "shock" and "dismay" "absolutely does" describe his reaction to the board's decision (author's interview with Gillum, 14 September 1994). Daniel Duncan, on

the other hand, said that this is an inaccurate description of his response; he stated that the judgment was in fact not at all surprising to him and was even anticipated (phone conversation with Duncan, 5 January 1995).

No licensing board or ethics committee has been able to give me an instance of a revocation for negligence prior to the Hyatt case. Stephen Unger (in *Controlling Technology: Ethics and the Responsible Engineer*, 2nd ed. [New York: Holt, Rinehart and Winston, 1994], esp. pp. 167–202) states that he, too, has been unable to find evidence of an engineer's being disciplined for endangering the public safety. George M. Bell, in his article "Professional Negligence of Architects and Engineers," *Vanderbilt Law Review* 12 (1959): 711, stated that "there has been no case which has ruled the architect or engineer liable for a negligently caused physical injury to persons on or off the premises after the structure has been turned over to the owner."

36. The most accessible description of the design change I have found is the rope analogy, for which I am indebted to Henry Petroski, *To Engineer Is Human: The Role of Failure in Successful Design* (New York: Random House, 1982), p. 88. Information on the time required for others to determine the cause is derived from my interview with Jack Gillum, 14 September 1994, as well as "Critical Design Change Is Linked to Collapse of Hyatt's Sky Walks," *Kansas City Star*, 21 July 1981, p. 1A; and Henry Petroski, *Design Paradigms: Case Histories of Error and Judgment in Engineering* (Cambridge: Cambridge University Press, 1994), p. 61.

37. The authors of the NBS report made it clear that their purpose was not to assign responsibility: see, for example, *NBS Report*, p. 2. Daniel Duncan confirmed this in a telephone interview on 5 January 1995. Paul Munger commented on the licensing board's misunderstanding as to the role of the NBS in my interview with him on 13 July 1994. Robert C. Flory described the ASCE decision to await the board's actions in my interview with him on 15 July 1994; also see American Society of Civil Engineers, Committee on Professional Conduct (ASCE-CPC), "Disciplinary Proceedings: Case of Mr. Jack D. Gillum," Docket No. 1982–6, 16 July 1986. A copy of the ASCE-CPC proceedings was provided by James W. Gillespie, the ASCE-CPC member who headed the Gillum case; my thanks to him both for providing the document to me and for obtaining permission for me to have access to it. Information on the board's discretion was obtained in interviews with Paul Munger, 13 July 1994 and 19 July 1994; with Robert C. Flory, 15 July 1994; and with Paul Spinden (assistant attorney general

during the Hyatt case), 5 July 1994. For an analysis of the media's coverage, see "A Case Study in Mass Disaster Litigation," *University of Missouri–Kansas City (UMKC) Law Review* 52 (1984): 174–75. Paul Munger stated that the results of the board's investigation "forced" the board "into making the determination" as to responsibility.

38. Discretion prevents me from reprinting here a sample of the grisly accounts, but a *Kansas City Star* article on 20 July 1981 ("Rescuers Rest, But Wrestle with the Horror," p. 2A, cols. 3–4) contains an example of the details provided. Minute-by-minute accounts of the evening were not uncommon in newspaper coverage and also appeared in the *NBS Report*. The *Kansas City Times* provided in-depth coverage of four families touched by the tragedy, listing what victims were wearing and even what they had eaten for breakfast that morning, in a special section entitled "In Memoriam, July 17, 1981, The Day the Music Stopped" (25 July 1981, sec. H).

39. Diane Stafford, "Grief Lies Heavily on Family and Firm with a Vision," *Kansas City Star*, 19 July 1981, p. 8HR; John T. Dauner, "Donald Hall Hopes to Leave Sorrow Behind," *Kansas City Times*, 20 July 1981, p. A4; John A. Dvorak, "Questions Loom as Hotel Designers Search for Cause," *Kansas City Times*, 20 July 1981, p. A1 (see especially the chart on p. A4); and George Koppe, "Portraits of the Firms That Built the Hyatt: Who's Who among the Principals," *Kansas City Star*, 26 July 1981, p. 29A (including a chart, "From Top to Bottom, the Firms Behind the Skywalks").

40. The chain of events has never been exactly determined, but two possible scenarios have been proposed in studies of the Hyatt. One possibility is that the fabricators were unable to obtain rods long enough to construct the walkways with single rods. The other scenario is that the construction team realized that threading 30-foot-long rods for the required nuts would be impractical, if not unsafe. See James B. Deutsch, "Statement of the Case," *Missouri Board v. Duncan, Gillum, and GCE International*, Case AZ–84–0239, filed 15 November 1985; *NBS Report*; "Hyatt Hearing Traces Design Chain," *Engineering News-Record* 213 (26 July 1984): 12–13; "Designers of Hyatt Hotel on Trial," *Civil Engineering* 54 (September 1984): 12–14; George F. W. Hauck, "Hyatt Regency Walkway Collapse: Design Alternatives," *Journal of Structural Engineering* 109 (May 1993): 1226–34.

41. For the evidence against Duncan and Gillum, see ASCE-CPC, "Disciplinary Proceedings." Duncan described the confusion resulting from

his exchanges with the fabricators and others in a phone conversation with me on 5 January 1995.

42. See John D. Constance, *How to Become a Professional Engineer: The Road to Registration* (New York: McGraw-Hill, 1988), p. 29, for history of Missouri licensing board. The newsletter of the Missouri Board, *Missouri Dimensions*, provided a statement of "methods pursued" (6 [March 1993]: 7, and also in other issues), as well as descriptions entitled "The Complaint Process—What Happens after Filing" (7 [June 1993]: 4) and "Complaint Handling and Disposition Procedure" (2 [May 1991]: 6–7). The procedures that minimize disciplinary actions were confirmed in a telephone conversation with Judy Kempker, administrative assistant for the board, on 27 October 1993 and in a telephone conversation with Tracy Hearst, secretary for the board, on 27 October 1993. See also Deutsch, "Statement of the Case," p. 12, where Judge Deutsch states that "the purpose of professional licensing is to protect the public, and not necessarily to punish the offender." Missouri Revised Statutes 327.411 RSMo 1978 state that a professional engineer is responsible for the content of plans on which he or she places a seal.

43. Deutsch, "Statement of the Case," pp. 243, 246; Missouri Revised Statutes, Chapter 327 RSMo 1978; "Petition for Reinstatement," Exhibit B, "Hyatt Regency Collapse Excerpts from Findings of State of Missouri Administrative Hearing," p. 2. The *New York Times* also reported that Duncan and Gillum "had told the board the accident was the result of poor communication" ("Panel Revokes Licenses of Two Hyatt Engineers," *New York Times*, 23 January 1986, p. I12). In interviews Jack Gillum and Daniel Duncan both confirmed that they felt lack of communication played a major role in the failure (author's interview with Gillum, 14 September 1994; author's phone conversation with Duncan, 5 January 1995). Louis L. Bucciarelli has described how the array of people and interests involved in the engineering design process affect the ultimate product in *Designing Engineers* (Cambridge: MIT Press, 1994).

44. Author's interviews with Jack D. Gillum, 31 January, 15 June, and 17 August 1994; Gillum, "Statement to the Board," Petition for Reinstatement, p. 7 (Gillum papers in author's files).

45. Claudia A. Cositore, ASCE-CPC secretary, to Jack D. Gillum, 31 May 1983, in ASCE-CPC, "Disciplinary Proceedings," Exhibit 3.

46. ASCE-CPC, "Disciplinary Proceedings," Exhibit 1, p. 2, and "Closing Statement." Also, author's interviews with James W. Gillespie, 22

June and 12 July 1994; Paul Munger, 13 July and 19 July 1994; and Grace Wald-Vogel at the ASCE throughout the spring and summer of 1994; Jack D. Gillum, "Statement to the Board," part of "Petition for Reinstatement," p. 7; and Greg Luth, "Hyatt Regency Collapse, July 17, 1981, Chronology of Events," exhibit B of "Petition for Reinstatement," p. 3. Consult "Disciplinary Matters," *ASCE News*, August 1986, and "Board Takes Disciplinary Action," *Civil Engineering* (ASCE) 56 (September 1986), for reports of the hearing (copies of both were provided by Grace Wald-Vogel, ASCE's senior manager for professional activities).

47. "New York Engineers Assail Ruling on Hyatt Regency Collapse," *New York Construction News* 33 (n.d.): 1, 11 (copy in Gillum papers in author's files); "Hyatt Ruling Rocks Engineers," *Engineering News-Record* 214 (28 November 1985): 13; Letters to the Editor, *Engineering News-Record* 216 (23 January 1986): 9–20; Robert A. Rubin and Lisa A. Banick, "The Hyatt Regency Decision: One View," *Journal of Performance of Constructed Facilities* 1 (August 1987): 161–67; Robert A. Rubin, Lisa A. Banick, and C. H. Thornton, "The Hyatt Decision: Two Opinions," *Civil Engineering* (ASCE) 56 (September 1986): 69–72; fourteen letters to the California Board of Registration for Professional Engineers by supporters of Jack Gillum on the occasion of his petition for reinstatement (copies in Gillum papers); Petroski, *To Engineer Is Human*, esp. pp. 90–91; and author's interviews with Paul Munger, 13 July and 19 July 1994; James Gillespie, 22 June and 12 July 1994; and Hon. Paul M. Spinden, 5 July 1994.

48. Charles Perrow, *Normal Accidents: Living with High-Risk Technologies* (Princeton, NJ: Princeton University Press, 1999).

49. Ibid., p. 5. Jack Gillum and Daniel Duncan have both attempted to counter this view of the inevitability of failures in their work since the Hyatt disaster. Since 1986, both have written and presented papers on recommended changes for the profession. Author's interview with Jack Gillum, 14 September 1994; "Petition for Reinstatement," Narrative Statement; author's phone conversation with Daniel Duncan, 5 January 1995.

50. Perrow, *Normal Accidents*, p. 7.

Three: "A New Era"

Epigraphs: Klemencic quoted by Nadine M. Post, "After Disaster: The Three R's of Threat Design—Resist, Respond, Recover—Are Catapulted to the Front Lines," *Engineering News-Record* 247 (31 December 2001): 8; "Continuing Lessons of 9/11," *New York Times*, 20 May 2004, p. A26.

1. "NIST Releases Findings on Collapse of Twin Towers," *ASCE News* 30 (May 2005): 7; Council on Tall Buildings and Urban Habitat, "Task Force on Tall Buildings: 'The Future,'" 15 October 2001, p. 31 (hereafter cited as CTBUH), www.ctbuh.org.

2. Post, "After Disaster," p. 8; "Continuing Lessons of 9/11," p. 26. Klemencic was chairman of the Council on Tall Buildings and Urban Habitat, an international nonprofit group dedicated to studying the impact of tall buildings on the environment. Thomas Kuhn, in *The Structure of Scientific Revolutions* (Chicago: University of Chicago Press, 1962), argued that a major "paradigm shift" in scientific understanding and theory required that scholars reach a crisis point sufficient to force them to shift to a new way of thinking. Only when presented with data or circumstances that could in no way be explained by existing theory would the scientific community be able to fully explore and adopt a new theory.

3. Henry Petroski, "The Fall of Skyscrapers," *American Scientist* 90 (January–February 2002): 16–20; quotation from p. 16.

4. Post, "After Disaster," p. 8.

5. CTBUH, pp. 47, 49.

6. Ibid., pp. 12–13; see also p. 53.

7. Ibid., pp. 21, 33–34; Post, "After Disaster," p. 8.

8. CTBUH, p. 40, 43.

9. Ibid., p. 40.

10. Ibid., pp. 43, 13, 44; Post, "After Disaster," p. 8.

11. CTBUH, pp. 44, 47.

12. Ibid., pp. 21, 38.

13. Richard Weingardt, "Hollywood's Structural Engineer," *Structural Engineer*, August 2002, p. 14.

14. In the process, it would also be important that the "solution" not make matters worse. See, for example, Nick Madigan, "$9.6 Billion Plan Announced for Redesign of Los Angeles Airport to Thwart Terrorists," *New York Times*, 10 July 2003, p. A18. Part of the plan called for a new facility dedicated to check-ins, which a RAND Corporation study criticized because "concentrating all passengers in the new check-in location . . . would actually increase the number of casualties in a terrorist attack."

15. Kenneth L. Carper, ed., *Forensic Engineering: Learning from Failures* (New York: ASCE, 1986), pp. 19–20.

16. Eric Klinenberg, *Heat Wave: A Social Autopsy of Disaster in Chicago* (Chicago: University of Chicago Press, 2002), pp. 9–11, 13.

17. Charles Fleddermann, *Engineering Ethics*, 3rd ed. (Upper Saddle River, NJ: Prentice Hall, 1999), p. 72. Interestingly, Fleddermann draws on none other than William Langewiesche for this classification scheme: see Langewiesche, "The Lessons of ValuJet 592," *Atlantic Monthly*, March 1998, pp. 81–98. A variety of other classifications of failures exist. See, for example, P. Aarne Vesilind, "Engineering and the Threat of Terrorism," *Journal of Professional Issues in Engineering Education and Practice* 129 (April 2003): 70–74; Kenneth L. Carper, ed., *Forensic Engineering* (New York: Elsevier, 1989), pp. 20–21; and James B. Eads et al., "On the Means of Averting Bridge Accidents," *ASCE Transactions* 4 (1875): 122–35.

18. J. H. Fielder and D. Birsch, *The DC-10 Case: A Study in Applied Ethics, Technology, and Society* (Albany, NY: State University of New York Press, 1992), e.g., pp. 1–12 and 69–81. See also Fleddermann, *Engineering Ethics*, pp. 84–85. The DC-10 design flaws were not limited to the cargo doors. These two books provide a description of the engine mount problems that also had disastrous consequences.

19. Steven Casey, *Set Phasers on Stun and Other True Tales of Design, Technology, and Human* Error, 2nd ed. (Santa Barbara, CA: Aegean Publishing, 1998), pp. 37–38.

20. Cascy, *Set Phasers on Stun*, p. 39.

21. Fleddermann, *Engineering Ethics*, pp. 67–93; quotation on p. 77.

22. For more on idiot-proofing designs, see Louis L. Bucciarelli, "Is Idiot Proof Safe Enough?" in *Ethics and Risk Management in Engineering*, ed. A. Flores (New York: University Press of America, 1989).

23. I coined the phrase "intentional accidents" in "Learning from Failure: Terrorism and Ethics in Engineering Education," feature article, *IEEE Technology & Society Magazine*, 21 (Summer 2002): 21, to describe engineering disasters resulting from intentional acts.

24. Vesilind, "Engineering and the Threat of Terrorism," p. 72.

25. Sherie Winston, "Wedge Renovations Are Charging Ahead," *Engineering News Record* 249 (2 September 2002): 9; Winston, "Pentagon

Rehab Pushes Innovation," *Engineering News Record* 249 (8 July 2002): 10.

26. See, for example, David W. Dunlap, "Manhattan: A Post-Sept. 11 Laboratory in High-Rise Safety," *New York Times*, 29 January 2003, p. C4, describing the new WTC7 design and how it incorporates lessons from 9/11. On the lessons learned from Galloping Gertie, see, for example, Richard Scott, *In the Wake of Tacoma: Suspension Bridges and the Quest for Aerodynamic Stability* (Reston, VA: ASCE, 2001). Henry Petroski has written extensively on how engineering science has evolved and improved over the past 2,000 years thanks largely to the study of failures; see, for example, *Success through Failure: The Paradox of Design* (Princeton, NJ: Princeton University Press, 2008). The ASCE study is Robin Shepherd and J. David Frost, eds., *Failures in Civil Engineering: Structural, Foundation, and Geoenvironmental Case Studies* (New York: ASCE, 1995).

27. Post, "After Disaster," p. 8. See also Christopher M. Foley, "Why They Fell," *ASEE Prism*, 11 (December 2001): p. 7.

28. Jim Dwyer and Kevin Flynn, *102 Minutes: The Untold Story of the Fight to Survive inside the Twin Towers* (New York: Times Books, 2005), pp. 65–69, 117–18. Dwyer and Flynn point out that the Port Authority was not required to follow the New York City building code or any other building code, but chose to do so. They also note, however, that the building code of 1968 was laxer than its predecessor, which had in part grown out of lessons learned from the deadly Triangle Shirtwaist Factory fire earlier in the century. In particular, "plans did not envision thousands of people weaving down the three staircases in each tower" (p. 168). Granted, "in comparison with 1993, when the evacuation down the dark and smoky stairs had dragged on for hours, from noontime into night, this one moved at lightning pace" (p. 170). Improvements in the concourse running underneath the towers (also made after the 1993 bombing) aided in the evacuation as well (p. 173). Roof rescues were not planned for: "Most experts did not view the roof as a viable escape route from a high-rise fire. For one thing, confusion about which way to walk on the stairs could cripple a large-scale evacuation. And the only way to rescue people from the roof was by helicopter, a method that could not be counted on to evacuate masses of people, certainly not anywhere near the number that a single intact staircase could accommodate" (p. 129). Moreover, "the Port Authority did not explicitly tell the occupants of its towers that the roof was not an

option" (p. 129). Tenants "faced the brutal truth that all the planners and drills had evaded. They had nowhere else to go" (p. 129).

29. Information on the 1975 fire is in National Institute of Standards and Technology (NIST), *Final Report on the Collapse of the World Trade Center Towers* (Washington, DC: U.S. Department of Commerce, 2005), p. 91.

30. Dwyer and Flynn, *102 Minutes*, pp. 49–51.

31. Ibid., esp. pp. 53, 55, 57, 60–61, 213–15, 223, 227, 247–48.

32. Ibid., pp. xxii, xxiii, 40, 80, and 243. Contrast, for example, James Glanz and Eric Lipton, "Expert Report Disputes US on Trade Center Collapse," *New York Times*, 22 October 2002, p. B1, quoting an unidentified source familiar with the Silverstein report: "The terrorists were the murderers here, not the buildings."

33. On lifeboat provisions in particular, see Wyn Craig Wade, *The Titanic: End of a Dream*, (New York: Penguin, 1986), pp. 302, 306–7. Also see Joel Baglole, "Iceberg Cowboys: Protecting Oil Rigs from *Titanic*'s Fate," *Wall Street Journal*, 16 May 2002, pp. A1, A10.

34. Wade, *The Titanic*, p. 321.

35. Ibid., p. 42.

36. See, for example, James Glanz and Eric Lipton, *City in the Sky: The Rise and Fall of the World Trade Center* (New York: Times Books, 2003), pp. 134–35.

37. Louis L. Bucciarelli, *Designing Engineers* (Cambridge, MA: MIT Press, 1994), p. 48.

38. Weingardt, "Hollywood's Structural Engineer," p. 14. This faith in engineering's ability to address disasters is evident in one of FEMA's recommendations: when the report notes that safety would be enhanced by cooperation and communication across fields, the investigators are referring to fields largely within structural and construction engineering.

39. "Testimony of Dr. W. Gene Corley on Behalf of the American Society of Civil Engineers before the Subcommittee on Environment, Technology and Standards & Subcommittee on Research Committee on Science U.S. House of Representatives March 6, 2002," p. 7, at www.asce.org/pdf/3-6-02wtc_testimony.pdf.

40. See, for example, Norbert Joseph Delatte, *Beyond Failure: Forensic Case Studies for Civil Engineers* (Reston, VA: ASCE Press, 2008); James Chiles, *Inviting Disaster: Lessons from the Edge of Technology* (New York: Harper Collins, 2001); or Susan Davis Herring, *From the* Titanic *to the*

Challenger: *An Annotated Bibliography on Technological Failures of the Twentieth Century* (New York: Garland, 1989).

Four: "Safe from Every Possible Event"

Epigraph: Leslie E. Robertson, "Reflections on the World Trade Center," *The Bridge* (National Academy of Engineering) 32 (Spring 2002): 10.

1. See John Seabrook, "The Tower Builder," *New Yorker*, 19 November 2001, 64–73, for more information on such tradeoffs.

2. Robertson, "Reflections on the World Trade Center," p. 10.

3. James Glanz and Eric Lipton, *City in the Sky: The Rise and Fall of the World Trade Center* (New York: Times Books, 2003), pp. 37 (on location), 60 (on decision to go to 10 million square feet of space), 106 (on tallest building demand), and 104 (on cost); Angus Kress Gillespie, *Twin Towers: The Life of New York City's World Trade Center*, rev. ed. (New York: New American Library, 2002), pp. 67–69 (on infill) and 76 (on typical costs).

4. Seabrook, "The Tower Builder," p. 67.

5. Seabrook, p. 73, notes that the Petronas Towers did in fact use a concrete core and concrete columns, though this structure was built in a different era than the Twin Towers.

6. Jim Dwyer and Kevin Flynn, *102 Minutes: The Untold Story of the Fight to Survive inside the Twin Towers* (New York: Times Books, 2005), p. 75 (see also pp. 154–56, 162–63); Glanz and Lipton, *City in the Sky*, p. 108. See also Gillespie, *Twin Towers*, pp. 80–84.

7. See Seabrook, "The Tower Builder," p. 67, on how "the same innovations that make these buildings more economical to erect and more pleasant to inhabit also make them more vulnerable to fire." On p. 73, Seabrook discusses how refuge floors provide either safe haven or a risky temptation, depending on the circumstances of the emergency. And finally, on p. 68 Seabrook talks of how difficult it is to assess whether a similar terrorist attack on an older building would have been more or less destructive.

8. Ibid., p. 73. Such trade-offs are not limited to skyscrapers. Consider, for example, developments in automobile design that make cars safer in crashes but more dangerous in rescues (Tim Moran, "What Makes You Safer in a Crash May Pose Risks in the Rescue," *New York Times*, 29 August 2005, p. 10).

9. Seabrook, "The Tower Builder," p. 64.

10. Glenn Fowler, "Broad Revisions of Building Code Proposed to City: Changes Would Allow Wider Freedom in Architecture—Costs Would Be Cut," *New York Times*, 9 July 1965, cited in Dwyer and Flynn, *102 Minutes*, p. 106. Ironically, the NIST final report recommended that post-9/11 safety would be improved by a building code that emphasized meeting certain performance criteria rather than on following specific building standards; see National Institute of Standards and Technology (NIST), *Final Report on the Collapse of the World Trade Center Towers* (Washington, DC: U.S. Department of Commerce, 2005), p. xlii.

11. Dwyer and Flynn, *102 Minutes*, p. 112.

12. Gillespie, *Twin Towers*, pp. 117–18.

13. NIST, *Final Report*, p. 19.

14. Richard Weingardt, "The Stewards of America's Structures," *Structural Engineer* 2 (January 2002): 16; Jon Magnusson, "Speak Up," *Civil Engineering* 74 (March 2004): 8; Seabrook, "The Tower Builder," p. 73; Christopher M. Foley, "Why They Fell," *ASEE Prism* 11 (December 2001): 6–7.

15. As in, for example, Henry Petroski, "The Success of Failure," *Technology and Culture* 42 (April 2001): 321–28.

Five: "Architectural Terrorism"

Epigraphs: John Seabrook, "The Tower Builder," *New Yorker*, 19 November 2001, p. 66; Body of Knowledge Committee of the Committee on Academic Prerequisites for Professional Practice, *Civil Engineering Body of Knowledge for the 21st Century* (Reston, VA: American Society of Civil Engineers, 2004), p. 12 (hereafter cited as *Body of Knowledge*).

1. Siva Vaidhyanathan, "An Entire Semester of Knowledge in One Day," *Chronicle of Higher Education*, 5 October 2001, p. B4.

2. Eric Darton, *Divided We Stand: A Biography of New York's World Trade Center* (New York: Basic Books, 1999), p. 119.

3. Eric Darton, "The Janus Face of Architectural Terrorism: Minoru Yamasaki, Mohammad Atta, and Our World Trade Center," in *After the World Trade Center: Rethinking New York City*, ed. Michael Sorkin and Sharon Zukin (New York: Routledge, 2002), pp. 88–89, 94.

4. Ibid., p. 90; "Continuing Lessons of 9/11," *New York Times*, 20 May 2004, p. A26. Darton was not alone in considering the connections

between architecture and terrorism; see, for example, Mark Wigley, "Insecurity by Design," in Sorkin and Zukin, *After the World Trade Center*, esp. p. 72, in which Wigley describes architects and terrorists not as twins but as mirror images, given their opposing interest in providing or removing a sense of security: "In this, the terrorist shares the expertise of the architect. The terrorist is the exact counter-figure to the architect." In *Longitudes and Attitudes: The World in the Age of Terrorism*, rev. ed. (New York: Anchor Books, 2003), pp. 126, 176, Thomas L. Friedman suggests that societies that support terrorism tend to be lacking in technological expertise and capabilities.

5. Eric Darton, "The Janus Face of Architectural Terrorism: Minoru Yamasaki, Mohammad Atta and the World Trade Center," 8 November 2001, at www.opendemocracy.net/conflict-us911/article_94.jsp. Also on the abstraction necessary to design, see Darton, *Divided We Stand*, p. 119.

6. Darton, *Divided We Stand*, p. 119; Darton, "The Janus Face" (2002), p. 90.

7. Samuel C. Florman, *The Existential Pleasures of Engineering*, 2nd ed. (New York: St. Martin's Griffin, 1994), pp. 137, 138.

8. Richard Feynman, *Surely You're Joking, Mr. Feynman*, ed. Edward Hutchings (New York: Norton, 1985), p. 137; emphasis in original.

9. Like Darton, *New Yorker* columnist Louis Menand equated the effects of terrorism and design: "We can say that the horror for the 4,000 victims of dying in those towers was inseparable from the horror of living in them." Louis Menand, "Faith, Hope and Clarity: September 11th and the American Soul," *New Yorker,* 16 September 2002, pp. 98–104, citing Jean Baudrillard's *The Spirit of Terrorism and Requiem for the Twin Towers,* trans. Chris Turner (New York: Norton, 2002).

10. Richard G. Weingardt, "The Three Approaches to Addressing Problems," www.eiass.com/BobProblems.htm; National Academy of Engineering, "Introduction to the Grand Challenges for Engineering," www.engineeringchallenges.org; Friedman, *Longitudes and Attitudes*, pp. 126 and 176. See also, for example, Samuel Florman, *The Introspective Engineer* (New York: St. Martin's Press, 1997), pp. 7 and 39.

11. Jim Dwyer & Kevin Flynn, *102 Minutes: The Untold Story of the Fight to Survive inside the Twin Towers* (New York: Times Books, 2005), p. 254.

12. *Faith and Doubt at Ground Zero*, videocassette, dir. Helen Whit-

ney (Frontline, PBS, 2002). For an example of technological fervor, see Florman's admission in *Existential Pleasures* of responding to critics with "perhaps more fervor than tact" and his attacks on what he described as anti-technologists' "personal phobia" (p. 68), their "bizarre nightmares" (p. 67), and their "poppycock" ideas (p. 117).

13. "Discover Dialogue: Anthropologist Scott Atran—The Surprises of Suicide Terrorism," *Discover Magazine* 24 (October 2003): 21–22 (see also extended version of this interview at www.discovermagazine.com); Scott Atran, "Genesis of Suicide Terrorism" *Science* 299 (7 March 2003): 1534–39 and supporting online material at www.sciencemag.org; Ariel Merari, "The Psychology of Extremism," paper presented to Institute for Social Research seminar series, University of Michigan, Ann Arbor, 11 February 2002, cited by Atran in both articles above.

14. National Commission on Terrorist Attacks, *The 9/11 Commission Report: Final Report of the National Commission on Terrorist Attacks Upon the United States* (New York: Norton, 2004), e.g., pp. 160–63 (quotation on p. 54). See also Friedman, *Longitudes and Attitudes*, p. 183, on the causes of terrorism.

15. Diego Gambetta and Steffen Hertog, "Engineers of Jihad," Sociology Working Papers, Paper 2007–10, Department of Sociology, University of Oxford, www.nuff.ox.ac.uk/users/gambetta/Engineers of Jihad.pdf, p. 2.

16. Ibid., pp. 6, 50; emphasis in original.

17. Jon Krakauer, *Under the Banner of Heaven: A Story of Violent Faith* (New York: Doubleday, 2003), p. 204, citing Anthony Storr, *Feet of Clay* (New York: Free Press, 1997), pp. 299, 300, 302.

18. Michael T. Gibbons, "Engineering by the Numbers," American Society of Engineering Education Report, posted at www.wepanknowledgecenter.org; and I. J. Busch-Vishniac and J. P. Jarosz, "Can Diversity in the Undergraduate Engineering Population Be Enhanced through Curricular Change?" *Journal of Women and Minorities in Science and Engineering* 10 (2004): 50.

19. Jeffrey Brainard, "50 Years after Sputnik, America Sees Itself in Another Science Race," *Chronicle of Higher Education*, 12 October 2007, p. A22. For an unusually balanced essay on this topic, see George Bugliarello, "Engineering, Foreign Policy, and Global Challenges" (editorial), *The Bridge* (National Academy of Engineering) 35 (Summer 2005), at www.nae.edu, Publications link.

20. *Body of Knowledge*, p. 12.

21. National Academy of Engineering, *The Bridge* 32 (Spring 2002), available at www.nae.edu, Publications link. See also the NAE's Engineer of 2020 page at www.nae.edu/Programs/Education/Activities10374/Engineerof2020.aspx.

22. ABET, "2009–2010 Criteria," www.abet.org/forms.shtml.

23. Nicholas Steneck, Barbara Olds, and Kathryn Neeley, "Recommendations for Liberal Education in Engineering: A White Paper from the Liberal Education Division of the American Society for Engineering Education," in *Proceedings of the 2002 American Society for Engineering Education Annual Conference*, CD-ROM (Washington, DC: ASEE), Session 1963, at www.asee.org.

24. Darton, "The Janus Face" (2002), p. 95.

Six: "These Material Things"

Epigraph: Leslie E. Robertson, "Reflections on the World Trade Center," *The Bridge* (National Academy of Engineering) 32 (Spring 2002): 9.

1. *World Trade Center: In Memoriam*, videocassette, prod. and dir. History Channel (A&E Home Video, 2002).

2. Ibid.

3. Robertson, "Reflections on the World Trade Center," p. 9; John Seabrook, "The Tower Builder," *New Yorker*, 19 November 2001, p. 66.

4. William Langewiesche, *American Ground: Unbuilding the World Trade Center* (New York: North Point Press, 2002), p. 63; Usha Lee McFarling, "Finding Hope in the Ruins," *Los Angeles Times*, 15 November 2001, p. A1.

5. Samuel C. Florman, *The Existential Pleasures of Engineering*, 2nd ed. (New York: St. Martin's Griffin, 1994), esp. pt. 3; quotation from p. 130.

6. Matthys Levy and Mario Salvadori, *Why Buildings Fall Down: How Structures Fail* (New York: Norton, 2002), pp. 13–14.

7. M. Christine Boyer "Mediations on a Wounded Skyline and Its Stratigraphies of Pain," in *After the World Trade Center: Rethinking New York City*, ed. Michael Sorkin and Sharon Zukin (New York: Routledge, 2002), p. 117 (referring to Michael J. Lewis's article "Before and After: In a Changing Skyline, a Sudden, Glaring Void," *New York Times*, 16 Sep-

tember 2001, p. 44); Moustafa Bayoumi, "Letter to a G-Man," in Sorkin and Zukin, *After the World Trade Center*, p. 132.

8. Mark Wigley, "Insecurity by Design," in Sorkin and Zukin, *After the World Trade Center*, p. 72; McFarling, "Finding Hope in the Ruins," p. A1.

9. Wigley, "Insecurity by Design," p. 70.

10. Langdon Winner, *The Whale and the Reactor: A Search for Limits in an Age of High Technology* (Chicago: University of Chicago Press, 1986). See also John Kuo Tchen, "Whose Downtown?!?" in Sorkin and Zukin, *After the World Trade Center*, esp. p. 37.

11. Raymond A. Mohl, *The New City: Urban America in the Industrial Age, 1860–1920* (Arlington Heights, IL: Harlan Davidson, 1985), pp. 46, 52.

12. Eric Darton, *Divided We Stand: A Biography of New York's World Trade Center* (New York: Basic Books, 1999), pp. 123–24, 146, 117.

13. For an accessible overview of the innovations present in the towers, see Robertson, "Reflections on the World Trade Center"; Angus Kress Gillespie, *Twin Towers: The Life of New York City's World Trade Center*, rev. ed. (New York: New American Library, 2002); or *World Trade Center: In Memoriam*. For more information on the origins of the towers, see especially James Glanz and Eric Lipton, *City in the Sky: The Rise and Fall of the World Trade Center* (New York: Times Books, 2003), pp. 33–34, 207, 215–17. Quotation from Wigley, "Insecurity by Design," p. 82.

14. Council on Tall Buildings and Urban Habitat, *Task Force on Tall Buildings: "The Future,"* 15 October 2001, pp. 35 and 40, available at www.ctbuh.org; Glanz and Lipton, *City in the Sky*, p. 217.

15. This often-cited quotation from Yamasaki appears, for example, in Ray Bernard, "WTC Profiles: Personal Stories of People Involved with the WTC Disaster," *Graduating Engineer and Computer Careers, 2002*, p. 24, available at www.graduatingengineer.com.

16. Sharon Zukin, "Our World Trade Center," in Sorkin and Zukin, *After the World Trade Center*, p. 13; Glanz and Lipton, *City in the Sky*, esp. pp. 62–87, 113, 173–74. See also Darton, *Divided We Stand*, pp. 130–43, 118; and Andrew Ross, "The Odor of Publicity" in Sorkin and Zukin, *After the World Trade Center*, pp. 121–23.

17. Celestine Bohlen, "The Old World under the New," *New York Times*, 18 August 2001, p. B7.

18. Glanz and Lipton, *City in the Sky*, p. 116, for the shipping boxes

comparison as well as other critiques; Lewis Mumford, *Pentagon of Power: The Myth of the Machine* (New York: Harcourt, Brace and World, 1970), 2:344, often misquoted as referring to "purposeless gigantism"; Hockenberry quote and Berman's response in Marshall Berman, "When Bad Buildings Happen to Good People," in Sorkin and Zukin, *After the World Trade Center*, p. 6.

19. On the symbolism, see, for example, Immanuel Wallerstein, "America and the World: The Twin Towers as Metaphor," unpublished manuscript online at http://fbc.binghamton.edu/iwbkln02.htm; Gillespie, *Twin Towers*, pp. 17–18; Lewis, "Before and After," p. 44.

20. Council on Tall Buildings and Urban Habitat, "Task Force on Tall Buildings," p. 39; Zukin, "Our World Trade Center," p. 13; David Harvey, "Cracks in the Edifice of the Empire State," in Sorkin and Zukin, *After the World Trade Center*, p. 57.

21. Bin Laden quoted in Daniel Benjamin, "The 1,776-Foot-Tall Target," *New York Times*, 23 March 2004, p. A23; *World Trade Center: In Memoriam.*

22. Seabrook, "The Tower Builder," p. 65.

23. On the "lukewarm prospective tenants," see Glanz and Lipton, *City in the Sky*, pp. 33–34. For other descriptions of the need (or lack thereof) for so much space, see Darton, *Divided We Stand*, pp. 123–24; Gillespie, *Twin Towers*, pp. 53–55, 188–207.

24. Christopher M. Foley, "Why They Fell," *Prism* 11 (December 2001): 7.

25. University Health Services, "Coping with Death, Grief, and Loss," University of Iowa, www.uiowa.edu/ucs/griefloss.html.

26. Florman, *Existential Pleasures*, p. 135.

27. Martin Heidegger, *The Question Concerning Technology and Other Essays*, trans. William Lovitt (New York: Garland, 1977), p. 4. This essay was originally published in 1954.

28. Robertson, "Reflections on the World Trade Center," pp. 9–10.

Conclusion: "More Time for the Dreaming"

Epigraph: Eric Darton, "The Janus Face of Architectural Terrorism: Minoru Yamasaki, Mohammad Atta, and Our World Trade Center," in *After the World Trade Center: Rethinking New York City*, ed. Michael Sorkin and Sharon Zukin (New York: Routledge, 2002), p. 95.

1. See, for instance, "The Fractured Landscape: Teaching, Reading, and Writing in the Wake of the September 11 Terrorist Attacks," *Chronicle of Higher Education*, 5 October 2001, pp. B4–B7.

2. Paul T. Durbin, "The Challenge of the Future for Engineering Educators," in *Ethics and Risk Management in Engineering*, ed. A. Flores (New York: University Press of America, 1989), pp. 233–36.

3. John D. Bransford, Ann L. Brown, and Rodney R. Cocking, eds., *How People Learn: Brain, Mind, Experience, and School* (Washington, DC: National Academy Press, 1999).

4. Ibid., especially chap. 2.

5. For a useful discussion of how to do so, see Daniel A. Vallero, "Teachable Moments and the Tyranny of the Syllabus: September 11 Case," *Journal of Professional Issues in Engineering Education and Practice* 129 (April 2003): 100–105. For other syllabi and lesson plans (predominantly non-engineering), see The Clarke Forum, "Teaching 9–11," Dickinson College, www.teaching9–11.org.

6. John Seabrook, "The Tower Builder," *New Yorker*, 19 November 2001, p. 73.

Recommended Reading

Those interested in reading more about the World Trade Center towers and the investigations into their collapse will find a rich collection of sources available. The following guide to a handful of the most useful accounts is arranged in roughly chronological order, from the creation of the World Trade Center to the completion of the investigations.

Angus Kress Gillespie and Eric Darton each wrote a biography of the towers before 9/11. Gillespie's *Twin Towers: The Life of New York City's World Trade Center*, rev. ed. (New York: New American Library, 2002), is a comprehensive and largely sympathetic account of the creation of the towers completed before September 11, accompanied by a video version produced by the History Channel entitled *World Trade Center: A Modern Marvel.* The book and video were both revised shortly after the collapse, with the video name changed to *World Trade Center: In Memoriam*, videocassette (A&E Home Video, 2002). Darton's *Divided We Stand: A Biography of New York's World Trade Center* (New York: Basic Books, 1999) is more critical of the towers and their creators and provides an instructive and thought-provoking counterpoint to Gillespie's version. James Glanz and Eric Lipton drew on their extensive reporting for the *New York Times* to write *City in the Sky: The Rise and Fall of the World Trade Center* (New York: Times Books, 2003), a balanced and careful look at the life of the towers from David Rockefeller's vision of a trade center at the heart of a revitalized Lower Manhattan to the stacks of twisted, salvaged steel lined up at the Fresh Kills landfill in Staten Island awaiting transformation into their next life.

On the morning of September 11, 2001, brothers Jules and Gedeon Naudet were in Lower Manhattan filming a documentary on a New York firefighter. Their camera caught footage of the first plane racing into the North Tower, and they were among the first to arrive on the scene just minutes later. The resulting video, *9/11*, dir. Jules Naudet, Gedeon Naudet, and James Hanlon (Paramount Video, 2001), is a remarkable and unique illustration of the shock and horror of that morning. *New York Times* journalists Jim Dwyer and Kevin Flynn meticulously compiled reports on many of the victims and survivors. Their book *102 Minutes: The Untold Story of the Fight to Survive inside the Twin Towers* (New York: Times Books, 2005) tells the story of 9/11 from the perspective of those inside the building and provides insight into how the occupants interacted with the buildings that day. William Langewiesche arrived on site shortly after the collapse and subsequently followed the progress of the "unbuilding" of the vast pile of rubble at Ground Zero in a series of *Atlantic Monthly* articles that later appeared as his book *American Ground: Unbuilding the World Trade Center* (New York: North Point Press, 2002). Additional eyewitness accounts are maintained in the September 11 Digital Archive at www.911digitalarchive.org.

In the weeks and months after the collapse, a variety of authors attempted to make sense of the collapse and to explore its implications for the future. Leslie Robertson wrestled with the loss of the buildings that he considered his children. John Seabrook visited with Robertson in these early days and reported on his encounter in "The Tower Builder," *New Yorker*, 19 November 2001, pp. 64–73. Several months later, Robertson described his reactions in a speech to his colleagues that appeared in print as "Reflections on the World Trade Center" in *The Bridge* (National Academy of Engineering) 32 (Spring 2002): 5–10.

College and university faculty members from around the country described the challenges of teaching about the collapse in a special section of the *Chronicle of Higher Education* in early October 2001 ("The Fractured Landscape," *Chronicle of Higher Education*, Chronicle Review Section, 5 October 2001). The Clarke Forum at Dickinson College established a website, www.teaching9-11.org, to provide a more comprehensive set of materials related to dealing with September 11 in the classroom. Michael Sorkin and Sharon Zukin compiled a provocative series of essays in their book *After the World Trade Center: Rethinking New York City* (New York: Routledge, 2002), which explore the meaning of the towers and their collapse for New York City and its residents.

Three government reports on 9/11 together provide a thorough and surprisingly accessible account of the events of that day. The report by the Federal Emergency Management Agency (FEMA) was the first to appear, in 2002, and is thus necessarily preliminary, but provides useful data and images and helpful descriptions; see Therese McAllister et al., *World Trade Center Building Performance Study: Data Collection, Preliminary Observations, and Recommendations,* FEMA Report 403 (Washington, DC: FEMA, Federal Insurance and Mitigation Administration, 2002). *The 9/11 Commission Report: Final Report of the National Commission on Terrorist Attacks upon the United States* (New York: Norton) appeared in 2004; it focused not on the mechanics of the collapse, but on providing detailed documentation of the terrorists and their actions leading up to 9/11. The National Institute of Standards and Technology (NIST) issued its *Final Report on the Collapse of the World Trade Center Towers* in 2005 (Washington, DC: U.S. Department of Commerce); this report is a more complete and conclusive version of the FEMA study. All three reports are readily accessible online.

Index